Mathematical Research · Mathematische Forschung

Wissenschaftliche Beiträge
herausgegeben von der
Akademie der Wissenschaften der DDR
Institut für Mathematik

Band 25

Parcella'84

Parcella'84

Proceedings of the II. International Workshop on Parallel Processing by Cellular Automata and Arrays held in Berlin (GDR), September 25–27, 1984

edited by

Wolfgang Händler
Tamás Legendi
Gottfried Wolf

Akademie-Verlag · Berlin 1985

Herausgeber:
Prof. Dr. Wolfgang Händler
Friedrich-Alexander-Universität
Erlangen-Nürnberg
Dr. Tamás Legendi
Hungarian Academy of Sciences
Research Group on Automata Theory,
Szeged
Dr. Gottfried Wolf
Akademie der Wissenschaften der DDR
Zentralinstitut für Kybernetik und Informationsprozesse, Berlin

Die Titel dieser Reihe werden vom Originalmanuskript
der Autoren reproduziert

ISSN 0138 - 3019

Erschienen im Akademie-Verlag Berlin, DDR-1086 Berlin,
Leipziger Straße 3-4

Lizenznummer: 202 · 100/409/85
Printed in the German Democratic Republic
Gesamtherstellung: VEB Kongreß- und Werbedruck, 9273 Oberlungwitz
Lektor: Dipl.-Phys. Gisela Lagowitz
LSV 1095
Bestellnummer: 763 471 3 (2182/25)
03000

PREFACE

This volume presents the results of Parcella '84 - the II. International Workshop "Parallel Processing by Cellular Automata and Arrays" held in Berlin from September 25 to 27 1984.
Like the previous workshop it was organized by T. Legendi from the Research Group on Automata Theory Szeged of the Hungarian Academy of Sciences and by G. Wolf from the Central Institute of Cybernetics and Information Processes, Berlin, Academy of Sciences of the G. D. R.
Mr. T. Legendi (Chairman), Szeged; G. Wolf (Co-Chairman), Berlin; V. Aladjev, Tallinn; P. E. Danielsson, Linköping; M. Gössel, Berlin; W. Händler, Erlangen; A. Jugel, Dresden; E. Katona, Szeged; H. T. Kung, Pittsburgh; J. Mikloško, Bratislava; H. Nishio, Kyoto, D. Parkinson, London; J. Prangishvili, Moscow; A. Salomaa, Turku; R. Vollmar, Braunschweig; G. Wunsch, Dresden were acting in the Programme Committee.

The Workshop united those scientists who were interested in discussing the new results and ideas for parallel and distributed processing on cellular and/or array systems. The existence side-by-side and the competition between these two closely related concepts influences their own development as well as the evolution of computer architecture in general. So the goal is to find more and deeper common aspects of these two system types and to stimulate this process.
The succesful work by Parcella '84 included not only the initiation of new ideas but also the exchange of information on people, places, efforts and objects of research.
The papers - especially of the supporting institutions - give a comprehensive overview about their results on this field.

The papers submitted were of good quality and the editors were highly at pains to select the most suitable ones. Following the idea to stimulate the mutual influence of cellular and array systems and because of the limitations put on the scope of the proceedings the editors were unable to accept several papers.
Furthermore they agreed that additional results of research areas not presented at the workshop might included in the proceedings.
The editors appreciated it very much that some authors could contritute their papers.

The Editorial Board would like to express its thanks to the members of the Programme Committee, the Chairmen of sessions and to all researchers who submitted their papers thus helping to achieve full success.

Finally the editors are very obliged to Prof. Dr. sc. V. Kempe, Director of the Central Institute of Cybernetics and Information Processes for his interest and for sponsoring and holding the workshop and to Dipl.-Phys. G. Lagowitz and Dr. R. Höppner from Akademie Verlag Berlin as publishing house for providing excellent conditions for edition of the proceedings.

Looking forward to the next Parcella '86

W. Händler

T. Legendi

G. Wolf

TABLE OF CONTENTS

+ not presented at the workshop

SESSION: ARRAY SYSTEMS

APPLICATIONS IMPLEMENTED ON THE ERLANGEN GENERAL PURPOSE ARRAY

W. Händler, G. Fritsch, J. Volkert

Universität Erlangen-Nürnberg
Institut für Mathematische Maschinen
und Datenverarbeitung (Informatik III)
Martensstraße 3, D-8520 Erlangen
Federal Republic of Germany

1. ABSTRACT

A processor array concept is presented which can be characterized by the following features:

- Homogeneity of the components: the processor-memory-modules are equal.
- Asynchronously operating processor-memory-modules.
- Communication between processors via memory.
- Restricted neighbourhood: each processor-memory-module is connected to a limited number of other modules.
- Organization of the modules in planes using a grid-like connection structure.
- Hierarchical arrangement of the planes such as to represent a "pyramid".

The multiprocessor operating system has a hierarchical structure corresponding to the hierarchical hardware structure. It resides in the upper planes whereas the application problems are processed by the worker processor array at the lowest level.

Due to the structure of the working plane various classes of algorithms for problems with an appropriate data structure are similar to those provided for cellular arrays.

Differences will be found in granularity. Such an algorithm, matrix multiplication, is presented. The implementation of this problem on the EGPA-pilot-system consisting of 4 worker processors resulted in a speedup of nearly 4. Furthermore, the implementation of a problem of non linear optimization is discussed.

On the basis of the results obtained from various implementations a larger EGPA-system is planned which will be used especially for multigrid-problems.

2. ARCHITECTURAL CONCEPT

Erlangen's concept for multiprocessors differs from others in some architectural respect. This is the reason for a success which mainly becomes evident in excellent speedups for a broad class of application areas (Fig.1).

Main points of an EGPA-system (or EGPA-structure) are [7, 8]:

1) Minimal complicacy (Fig. 2) is reached by using one unique type of elements (modules) - consisting of one processor and its allocated memory (not a private memory). The results are then:

   a) unique simple interface with respect to hardware and software

   b) simplified stockholding of spare-parts (modules of one unique type)

   c) simplified compensation of defects in runtime (mutual exchangeability of functions in modules) is possible (fault-tolerance concept).

2) Simplified coupling (Fig. 3) of modules by connecting processors to memories via multiports. Conflicts are resolved by the multiports without any care of users respectively programmers.

3) Asynchronous operation of modules (independently working besides in case of memory conflicts).

4) Communication between processors happens via (common) memories.

5) Restricted neighbourhood (Fig. 4) protects from a catastrophic broadcasting of failures created by a defective module. Each modul is connected to a restricted number of other modules only. Fault-tolerant behaviour can easier be realized than in a totally connected graph network .

6) Regular arrangements (Fig. 5) of module connections (rectangular and grid-like) offer the opportunity e.g. to solve partial differential equation by relaxation a more direct way (two-dimensional working area).

7) Hierarchical arrangements (tree-like) connecting the two-dimensional areas (according to 6)) to a pyramid (Fig. 6). The areas contain 1, 4, 16, 64 ... etc. processors (compare Fig. 5). The arrangement reflects - at the same time - the hierarchical structure of operating systems.

8) Size independent topology is a consequence of points 6 and 7. With respect to one processor the number of connections to memories never exceeds 9. Therefore pyramids of EGPA-topology can unrestrictedly grow - at least from a theoretical point of view. The theory is valid for thousands and even millions of modules (processor-memory module , PMM).

In the context of point 6 - regular arrangement (Fig. 5) of modular connections - is to state that the advantages hold with other applications

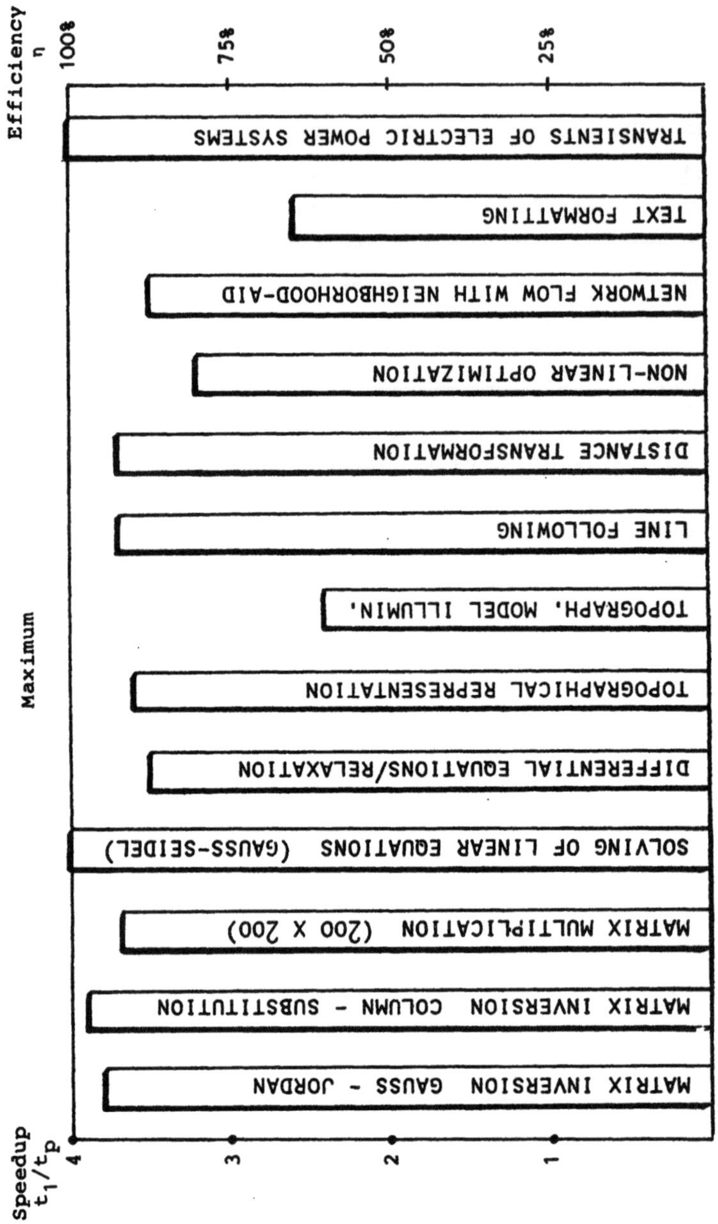

Fig. 1: Speedup (t_1/t_p) and efficiency ($\eta = (t_1/t_p)/p$) for a collection of applications

t_1 ≡ time for computation on a single processor

t_p ≡ time for computation on p processors

p ≡ number of processors working simultaneously

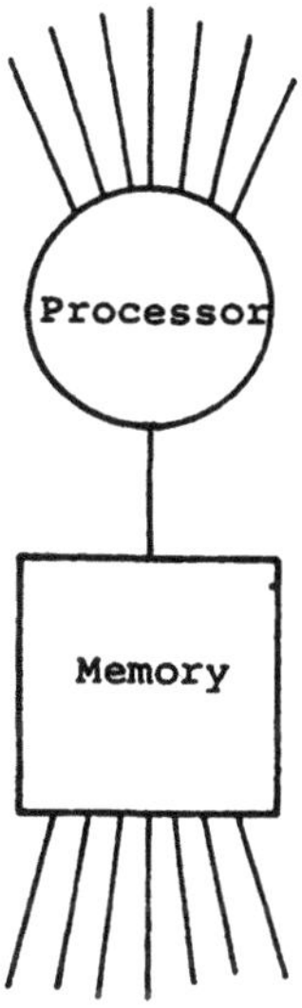

Fig. 2:
Processor-Memory-
Module

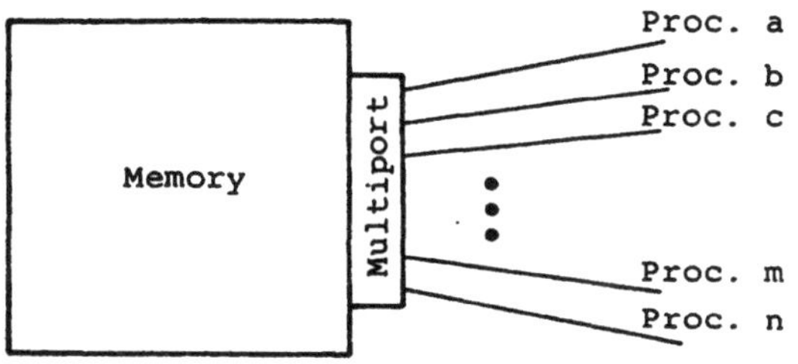

Fig. 3: Coupling by multiport
Conflict solving in the case of simultaneous demands from more than one processor.

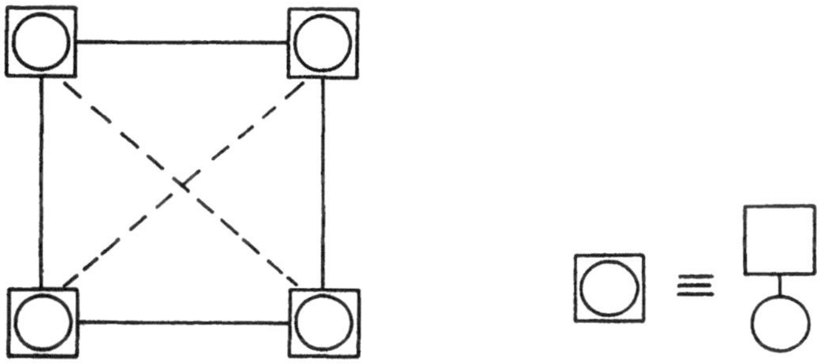

Fig. 4: Restricted neighbourhood (in principle).
Dashed lines are <u>not</u> realized (EGPA structure does not contain diagonal connections e.g.)

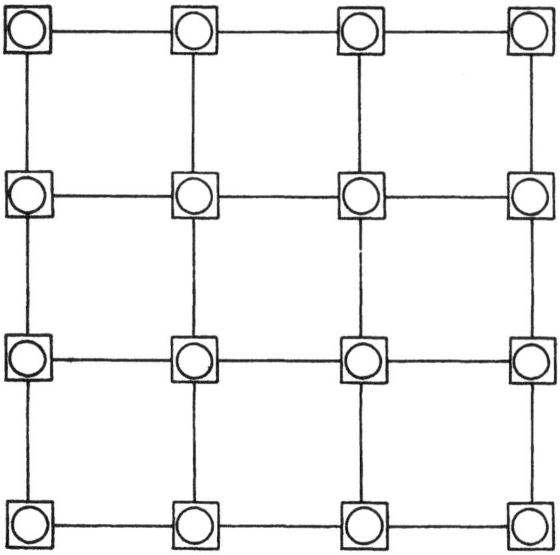

Fig. 5: Regular (discrete) topology of EGPA (two-dimensional working area).

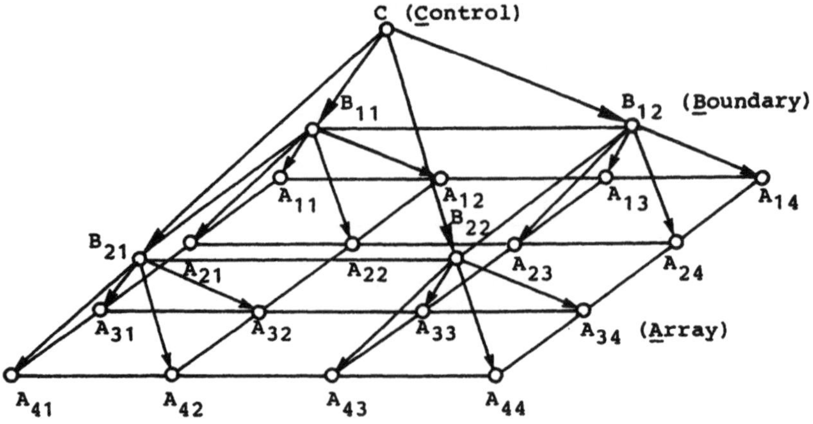

→ Unidirectional connection: processor - memory

— Bidirectional connection: processor - memory

O Processor and its associated memory

Fig. 6: EGPA-pyramid (two-dimensional working area - compare Fig. 5 - plus an additional, hierarchically organized dimension)

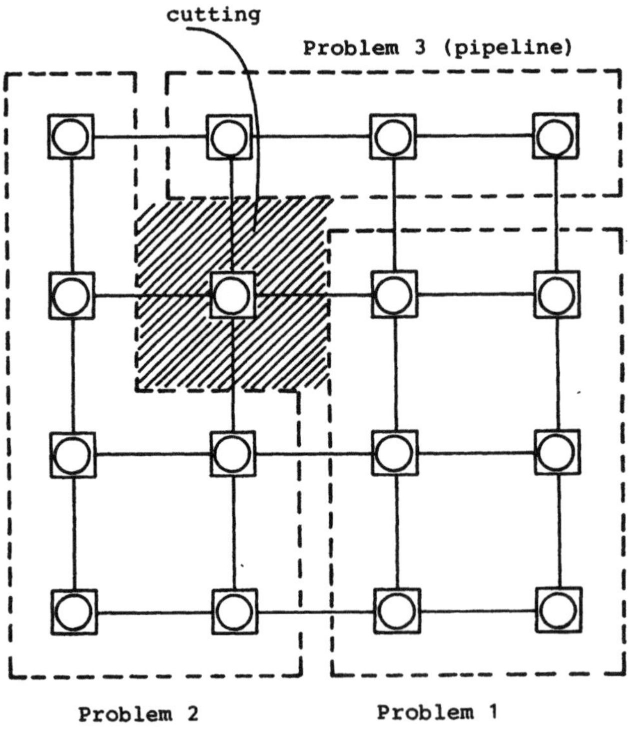

Fig. 7: Example of space-sharing (instead of time-sharing)

than partial differential equation. Indeed a certain evidence for those applications is given in which algorithms show an inherent regular structure. Nevertheless in numerous cases no connections at all or only a part of them are utilized. Then the rectangular array degrades to a stock of modules (processor/memory) for which some direct connections are provided. Seemingly irregular algorithmic structures - given by the necessity to exchange data from module to module - can be mapped onto or into a rectangular array as it is realized in EGPA-structures.

Is this mapping "into" then some modules are not used or at least not completely utilized in the case where other user problems respectively applications can be found which fill the array. Among many jobs a subset must be found for which a partition exists which can be mapped into or allocated to the array without too hard cuttings (Fig. 7).

Another case is if no mapping exists which ensures a direct data transfer (or exchange) between some of the concerned modules. Then two ways are open. On the one hand possibly one or more modules can be found which up to this point did not take part in the consideration and which now can play the role of a transfer-module. On the other hand a transfer can always be organized via the upper arrays of the pyramid (Fig. 8). In both cases an additional time for a pure transfer (redundant transfer with no processing else) is needed.

For a transfer of the latter kind the additional time lies in the range of logarithm of the "distance" in the array (working array). Distance is defined as the number of elementary transfer steps which in the array (in the plane) would have to be passed through (compare Fig. 8).

Experience at Erlangen shows that such a vastage of time can be avoided with careful planning. So it is the most awarding challenge to search for effective systematic methods for planning how a job stream can be allocated (or mapped) onto (or into) the working array (with a minimum of cuttings). Up to now case studies (compare next chapters) direct the intuitive way to solutions. A quite general method for bringing jobs to the multiprocessor-structure cannot be seen up to now and is even questionable.

At Erlangen therefore we followed a pragmatic way to work with many different applications in order to see what they have in common. The results are encouraging to gain more insight to the most exciting "parallelization" problems.

Assuming that an application is permanently repeated e.g. industrial control, traffic control, weather forecast etc. of a particular kind then the prefabricated structure of EGPA-systems are not well suited. Considerations of a taylored (and dedicated) multi-configuration have to be made then.

While an EGPA-structure is mainly suited to computing centres because of its general purpose nature, at Erlangen another concept is directed to a "construction box" (kit) DIRMU which enables the user or customer to arrange his own "multi" fully adapted to the need of a particular problem or a particular algorithm [15]. Adaption is possible by putting modules together via multiport (like in EGPA) in a nearly arbitrary way. Only one restriction is valid for DIRMU-configurations: the multiport of DIRMU-modules is - at least with our present pilot-implementation - laid out for 7 connections beyond the connection to the allocated module-memory. In the case where more than 7 external connections are needed, it is possible to cascade modules. The restriction is nevertheless reasonable not only because of hardware-expenses but also because of problems of traffic contention, fault-tolerance and deadlock-free- concepts.

The concept of DIRMU is not discussed in this paper beyond these brief remarks (compare [9]).

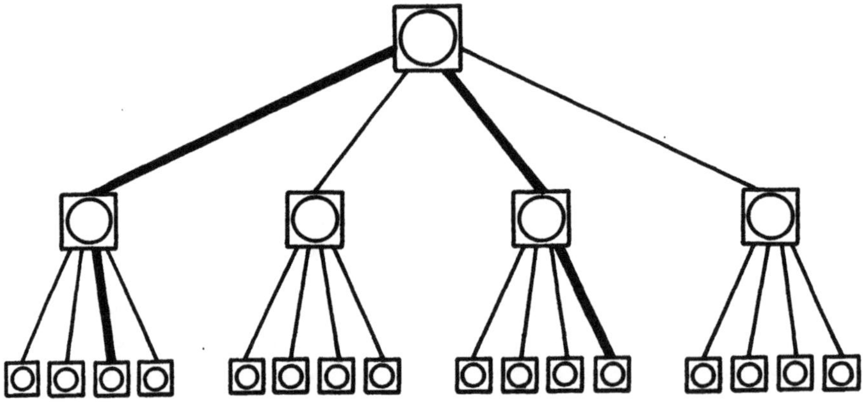

Fig. 8: Unwinding the EGPA-pyramid onto a plan and leaving out horizontal connections, arbitrary "connections" can be realized via the inherent tree-structure, which is given by the "vertical" dimension of the pyramid.
If d is the distance, which is defined as the number of connections data would have to traverse in the working area (plane), then the way through the upper regions of the pyramid amounts to

$$w = \log_2 d$$

elementary transfer periods.
The consideration shows at the same time that a subset of EGPA-pyramid-connections makes up a most comfortable tree-processor as developed elsewhere.
All problems suited to a tree-processor therefore can also be efficiently processed on an EGPA-structure.

3. APPLICATIONS

3.1 Survey

Parallel processing with the EGPA pilot system was tested with a number of quite different applications. The implemented problems and the obtained speedup values are summarized in the following table.

Linear algebra [5, 10]	SPEEDUP
- Matrix inversion	
Gauss-Jordan	3.8
Column-substitution	3.9
- Matrix multiplication	3.7
- Solving of linear equations	
Gaus-Seidel	4.0
Differential equations [4]	
- Relaxation	3.5
Image processing and graphics [6]	
- Topographical Representation	3.6
- Illumination of the topographical model	2.9
- Line following	
(vectorizing of a grey level matrix)	2.9
- Distance transformation	
each processor is working on a fixed part of data	3.0
dynamic assignment of varying parts of data	3.3
Non linear programming [3]	
- Search for minima of a multi-dimensional objective function given by an algebraic term	3.2
Graph theory	
- Network flow with neighbourhood-aid	3.5
Text formatting [17]	2.6
Maximal theoretical speedup (4 worker processors)	4.0

Table: Applications implemented on the pilot pyramid.

In the following chapter 3.2 two applications and the implementation results are discussed in detail. The first problem, a nonlinear optimization problem, belongs to a problem class with problem structures which do not easily allow matching the regular worker processor array. However, due to memory coupling between neighbouring processors this drawback of the problem structure can be smoothed by neighbourhood aid. The other application, a matrix multiplication problem, has a data structure similar to those provided for systolic arrays

3.2 A Non-linear Optimization Problem

3.2.1 General Remarks

In order to compute on a multiprocessor system a given problem must be decomposed into subtasks so as to match the multiprocessor structure. This "parallelization" work can be carried out at

- the proper problem level,
- the algorithm level (e.g. a numerical procedure) or at
- the programming level.

Frequently, parallelization at the algorithm level is most effective for computation on a multiprocessor system. However, parallelized versions of an algorithm can prove less efficient than the sequential one. This will be shown in section 3 with a minimum search problem.

On the other hand, decomposition into subtasks at the problem level mostly does not produce a computational structure which allows for a balanced load of the multiprocessor system. As an example, the analysis of an electrical power system may be taken.

One can split up the power network in several subnetworks, each one being computed on one worker processor, e.g. an A-processor of the EGPA-system. In order to minimize interprocessor communication, the number of connections between sub-networks must be minimized. This, in term, leads to a large variation of the number of nodes in the sub-networks which produces an inhomogeneous load distribution in the multiprocessor system. Therefore the decomposition is preferably effected at the algorithm level where matrix-vector-operations are being executed. By applying an appropriate decomposition strategy a homogeneous subtask structure can be obtained. With this application a speedup of the EGPA-system versus a monoprocessor of 4 was measured [19]

Parallelizing the given optimization problem merely at the program level did not result in an effective parallel program. This was mainly due to inherently sequential program portions.

3.2.2 Description of the user's problem

The user problem consists of the calculation of the reaction and transport parameters of a chemical reaction system. A least-squares function containing the sum of the squares of the differences between the experimental and the theoretical values of temperature and concentration is used as an objective function of the system. The theoretical values are calculated on the basis of a mathematical model for the real system to be studied. The model uses a set of parameters whose values can be determined by minimization of the objective function.

As a basis for an parallelization study we used a description of the mathemacial model of this chemical problem and its (sequential) FORTRAN program system FIBSAS [11] which has been tested on various computers (TR 440, SIEMENS PR 330, Cyber 172). Most computation time is needed for the minimum-search-subroutine. Therefore parallelization was concentrated onto the minimum-search-subtask, both at the algorithm and the programming level [3].

For minimization of the objective function the simplex-method of Nelder and Mead [16] was applied. This minimum search procedure is a stepping method: A simplex moves in the function landscape changing its shape by replacing the vertix with the highest function value by a new one with a lower function value. In this way the simplex proceeds towards the function minimum. The original procedure is sequential. It can be parallelized in various ways, e.g. by searching minima on different routes simultaneously.

3.2.3 Parallelization of the minimum search problem

Several parallel minimum-search strategies are possible from which an appropriate one has to be selected. A parallel version of the simplex algorithm should ensure that the assignment of the subtasks to the 4 A-processors (array processors) of the EGPA-system allows for an efficient computation. If a sequential algorithm runs on the monoprocessor (AEG 80-60) and a parallel algorithm runs on the multiprocessor system with the computation time T_1 and T_p respectively a speedup value

$$S = \frac{T_1}{T_p}$$

can be determined.

However, parallelization of the sequential simplex algorithm can diminish the computational efficiency. Moreover, the efficiency of a given parallel minimum-search-algorithm can strongly depend on the type of the objective function, in particular on the number and relative positions of the minima. Therefore, one needs some figure which represents a measure for the efficiency of the algorithm. The number of function evaluations n_s and n_p necessary to find a given minimum by use of a sequential and a parallel algorithm (program) respectively can be taken as a standard of comparison.

In order to take into account the difference between the efficiency of the sequential and that of the parallel algorithm the experimental speedup value should be multiplied by the factor n_p/n_s. Thus, the speedup of the multiprocessor with regard to the monoprocessor is obtained, the algorithm effect being eliminated.

We adopted two ways to run an appropriate minimum search algorithm on the EGPA-system:

(1) Design and implementation of a parallel version of the minimum search algorithm. For this purpose we modified the simplex algorithm so as to simultaneously compute the function values of four new candidates for simplex vertices (in accordance with the 4 worker processors of the EGPA pilot system).

(2) Subdivision of the minimum search area into as many subareas as worker processors are available. Each processor is allocated to one subarea and the sequential minimum search algorithm is applied.

According to the first way a parallel version of the minimum search algorithm was implemented on the EGPA-system. However, the results showed that the parallel minimum search algorithm was less efficient than the sequential one. This was due to the fact that the parallel algorithm proceeds less carefully, evaluating more values of the objective function (n_p) than the sequential algorithm (n_s). Therefore, we continued with the second way. Dependent upon the function landscape and the halting criterion the number of function evaluations can considerably vary from one subarea to the other. This can produce an unbalance of the computing load between the processors. The unbalanced load can be smoothed out by applying neighbourhood-aid that is using idle neighbouring worker processors to support still busy neighbours. This processor allocation strategy can efficiently be applied only with tightly coupled processor arrays.

3.2.4 Results

According to way (2) of the previous section 3 test-functions with 1, 4 and 11 minima were respectively implemented [18]. As an example the number of function evaluations (proportional to computation time) of the 4 worker processors (A-processors) in the case of the testfunction with 4 minima is given by

1503, 402, 354, 589.

This result demonstrates that a considerable unbalance of the load among the worker processors can occur. The load balance can be improved by application of neighbourhood-aid. The corresponding numbers are given in the next two lines which represent the results obtained with neighbourhood-aid of one and two processors respectively:

1003, 949, 354, 589
869, 848, 354, 859.

By this way a speedup of about 3 was obtained.

It can be expected that for larger EGPA-systems with 16 or 64 worker processors, each processor being connected to four neighbours, a still better balancing would be achieved.

The implementation of the FIBSAS-objective-function with 6 optimization parameters led to computation times between 8 and 24 hours. The computational procedure can be subdivided into 4 phases as explained in Fig.9 .

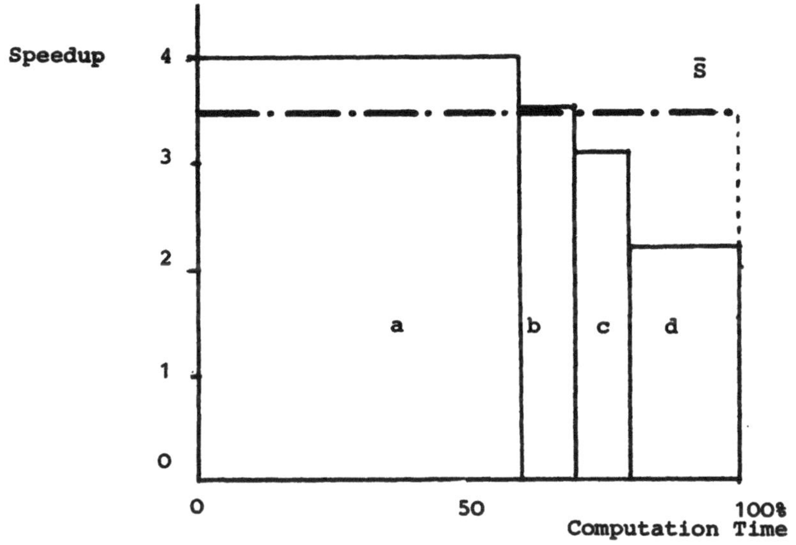

Fig. 9 : Computation of a minimum-search problem using the 4 worker processors (A-processors) of the EGPA-pilot-pyramid:

Phase a: All 4 A-processors are busy, each one over its subarea,

b: 3 A-processors are busy, one is effecting neighbourhood-aid,

c: 2 A-processors are busy, two are effecting neighbourhood-aid,

d: 1 A-processor is busy, two are effecting neighbourhood-aid, one is idle.

From the speedup of the four phases an averaged speedup $\bar{S}$ can be obtained

The speedup achieved with a real objective function depends on the number and species of the selected optimization parameters. In most cases a mean speedup $\bar{S}$ (compare with Fig. 9) between 3 and 3.5 can be obtained.

3.3 Matrix Multiplication

In the last paragraph we described one example of the class of algorithms which are not correlated with techniques used for cellular arrays. This fact is caused by the problem structure. It looks different for all methods which operate on matrix-like data. This especially includes many solutions of problem arising in picture processing and in linear algebra. One typical example is matrix multiplication.

3.3.1 Algorithm for a cellular space

We chose an algorithm introduced by E. Katona [12] who applied the cell-processor architecture of T. Legendi [14]. We cite E. Katona word-for-word:

The task is to compute the product $C = A \cdot B$ where A and B are <u>binary</u> matrices of size $n \cdot n$ (for rectangle-form matrices see point 3). The element c_{ij} of the product C is defined as $c_{ij} = a_{i1}b_{1j} + \dots + a_{in}b_{nj}$ where the multiplication means the <u>logical 'and'</u>, the addition means the <u>logical 'or'</u> operation. (This definition corresponds to the composition of relations, but the cellular algorithm works also in the case of other operations.)

To compute the product C a cellular space consisting of $n \cdot n$ cells will be applied. The individual state-bits of cells perform different functions:

- on the first (leftmost) bits of cells the elements of A move from left to right,
- on the second bits of cells the elements of B move from up to down,
- on the third bits of cells the elements of C will be computed,
- the fourth bits are not used.

The multiplication process is demonstrated on figure 10. In the initial configuration each cell is in state 0. The matrices A and B come from the outside world in slanting format (i.e. the matrix elements are written step by step in the left and upper dummy cells).

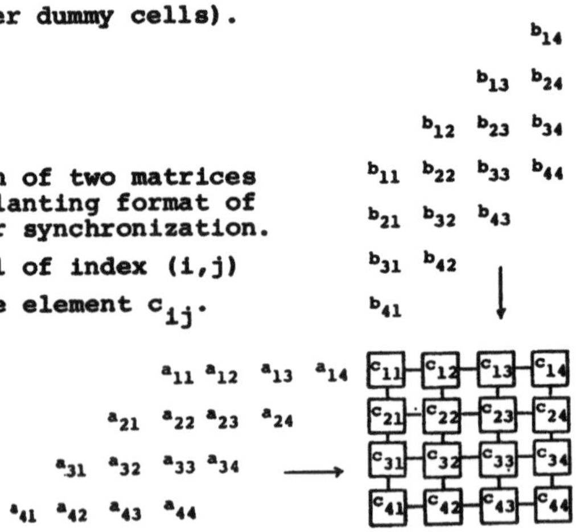

<u>Fig.10:</u> The multiplication of two matrices of size 4.4 The slanting format of A and B serves for synchronization. $\boxed{c_{ij}}$ means the cell of index (i,j) which computes the element c_{ij}.

It is obvious from the figure that the i-th row of A and the j-th column of B go through the cell of index (i,j) and the elements a_{ik} and b_{kj} meet in this cell for any k (synchronization). So the cell of index (i,j) can compute the element c_{ij} by the transition function

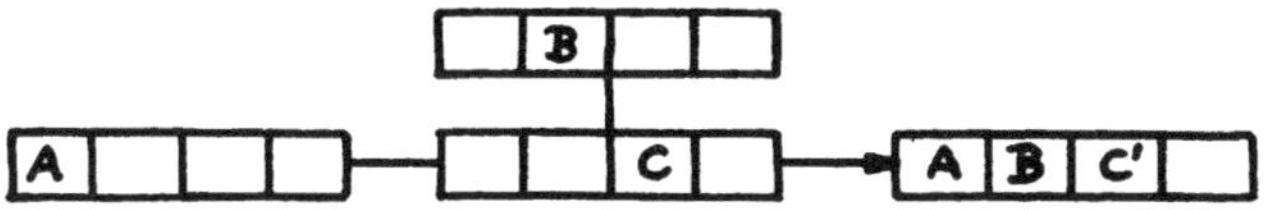

where C' = C + A · B

if each cell executes this transition function (homogeneity), then <u>after 3n steps</u> (when the matrices A and B leave the cellular space) the matrix C has been computed, the result can be found on the third bits of cells.

<u>3.3.2 Band matrix multiplication by systolic array</u>

Due to the great advances in large scale integration algorithms suitable to systolic arrays were created. Such an (two-dimenstional) array consists of processing elements (PE) arranged in a regular manner within a plane. Each PE is connected to some neighbours. These connections mostly are undirectional. The whole array forms a SIMD computer system: all PEs execute the same instruction at the time. This means that the array operates in a way equivalent to cellular space introduced in the last paragraph. But there is one essential difference: the basic data is the REAL and not the BOOLEAN.

The algorithm generates the product of two band matrices (see Fig. 11) and was introduced by Kung and Leiserson [13]

$$\underset{A}{\begin{bmatrix} a_{11} & a_{12} & & & 0 \\ a_{21} & a_{22} & a_{23} & & \\ a_{31} & a_{32} & a_{33} & a_{34} & \\ & a_{42} & & & \ddots \\ 0 & & & & \end{bmatrix}} \underset{B}{\begin{bmatrix} b_{11} & b_{12} & b_{13} & & 0 \\ b_{21} & b_{22} & b_{23} & b_{24} & \\ & b_{32} & b_{33} & b_{34} & b_{35} \\ & & b_{43} & & \ddots \\ 0 & & & & \end{bmatrix}} = \underset{C}{\begin{bmatrix} c_{11} & c_{12} & c_{13} & c_{14} & 0 \\ c_{21} & c_{22} & c_{23} & c_{24} & \\ c_{31} & c_{32} & c_{33} & c_{34} & \\ c_{41} & c_{42} & & & \ddots \\ 0 & & & & \end{bmatrix}}$$

<u>Fig. 11:</u> Band matrix multiplication

The following Fig. 12 describes the algorithm in a self-explaining manner. Therefore we do not go into details, but we emphasize that this method is obviously influenced by the theory on cellular arrays.

<u>3.3.3 Macro systolic matrix multiplication on an EGPA-system</u>

We describe a program for matrix multiplication H = F·G[10]. The essential idea has its origin in a paper of Cannon [1], but the simularity to the previous algorithms is evident. Thus, this is an example for the influence of research on cellular array to parallel programming of real computer systems.

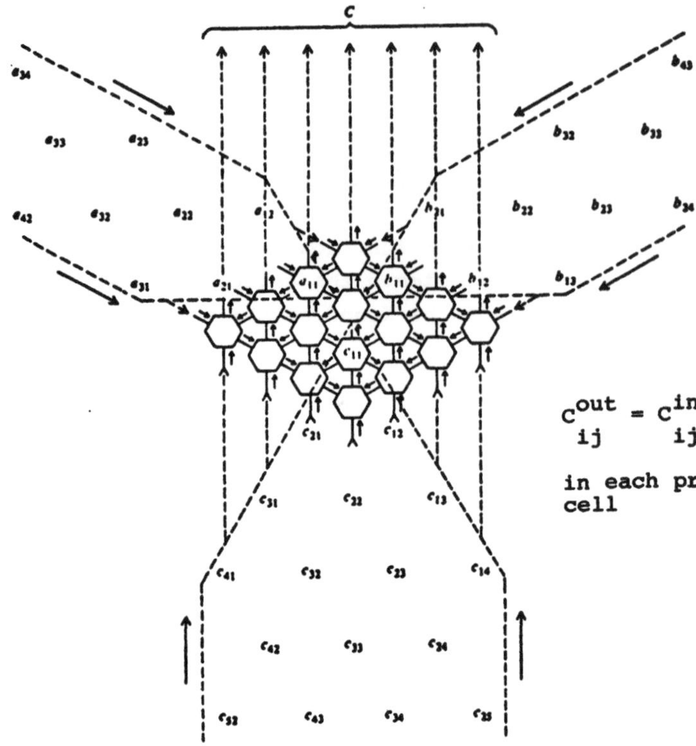

$c_{ij}^{out} = c_{ij}^{in} + a_{ik} b_{kj}$

in each processing cell

<u>Fig. 12:</u> A systolic array for band matrix multiplication

The algorithm is illustrated by Fig. 13 and has the following steps:

<u>Start:</u> We assume that at the beginning H contains only zeros and that all the matrices F, G and H are divided into rather equal portions which correspond to the structure of the working plane of our EGPA-system. The partial matrices $F_{i,j}$, $G_{i,j}$ and $H_{i,j}$ are stored in the memory of the processor $p_{i,j}$ in the row i and the column j of the lowest plane of the system (see Fig. 13 (Start), each square represents one PMM). The start position is tranformed by more or less rotation of the partial matrices $F_{i,j}$ and $G_{i,j}$ in rows respectively in columns until the data are distributed as depicted in Fig. 13 (Step 1). These data transports can easily be done by the hardware because of the toroidal connections.

<u>Step i:</u> Each processor $p_{i,j}$ multiplies the partial matrices $F_{i,j}$ and $G_{k,j}$ stored in its memory at the beginning of step i and adds the resulting matrix to $H_{i,j}$. Subsequently the processor writes the matrix $F_{i,k}$ into the memory of the processor "west" of it and the matrix $G_{k,j}$ into the memory of the processor "north" of it (for orientation see Fig. 13 (top)).

Result: After p steps (if there are p x p processors in the working plane), the result H is distributed on the array as shown in Fig. 13 (Result).

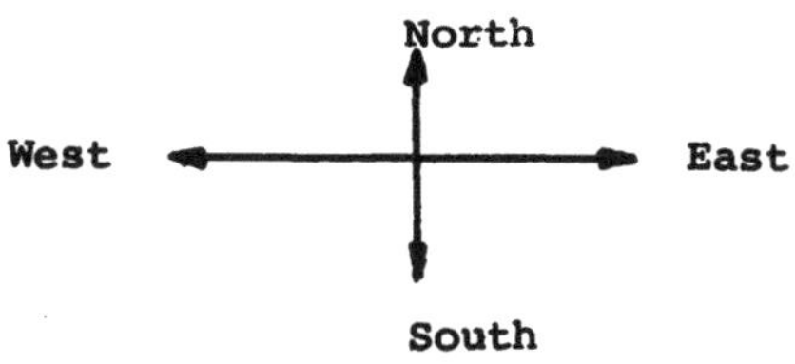

F_{11} G_{11}	F_{12} G_{12}	F_{13} G_{13}	F_{14} G_{14}
F_{21} G_{21}	F_{22} G_{22}	F_{23} G_{23}	F_{24} G_{24}
F_{31} G_{31}	F_{32} G_{32}	F_{33} G_{33}	F_{34} G_{34}
F_{41} G_{41}	F_{42} G_{42}	F_{43} G_{43}	F_{44} G_{44}

Start

F_{11} G_{11}	F_{12} G_{22}	F_{13} G_{33}	F_{14} G_{44}
F_{22} G_{21}	F_{23} G_{32}	F_{24} G_{43}	F_{21} G_{14}
F_{33} G_{31}	F_{34} G_{42}	F_{31} G_{13}	F_{32} G_{24}
F_{44} G_{41}	F_{41} G_{12}	F_{42} G_{23}	F_{43} G_{34}

Step 1

F_{12} G_{21}	F_{13} G_{32}	F_{14} G_{43}	F_{11} G_{14}
F_{23} G_{31}	F_{24} G_{42}	F_{21} G_{13}	F_{22} G_{24}
F_{34} G_{41}	F_{31} G_{12}	F_{32} G_{23}	F_{33} G_{34}
F_{41} G_{11}	F_{42} G_{22}	F_{43} G_{33}	F_{44} G_{44}

Step 2

F_{13} G_{31}	F_{14} G_{42}	F_{11} G_{13}	F_{12} G_{24}
F_{24} G_{41}	F_{21} G_{12}	F_{22} G_{23}	F_{23} G_{34}
F_{31} G_{11}	F_{32} G_{22}	F_{33} G_{33}	F_{34} G_{44}
F_{42} G_{21}	F_{43} G_{32}	F_{44} G_{43}	F_{41} G_{14}

Step 3

F_{14} G_{41}	F_{11} G_{12}	F_{12} G_{23}	F_{13} G_{34}
F_{21} G_{11}	F_{22} G_{22}	F_{23} G_{33}	F_{24} G_{44}
F_{32} G_{21}	F_{33} G_{32}	F_{34} G_{43}	F_{31} G_{14}
F_{43} G_{31}	F_{44} G_{42}	F_{41} G_{13}	F_{42} G_{24}

Step 4

H_{11}	H_{12}	H_{13}	H_{14}
H_{21}	H_{22}	H_{23}	H_{24}
H_{31}	H_{32}	H_{33}	H_{34}
H_{41}	H_{42}	H_{43}	H_{44}

Result

Fig. 13 : Shifting the partial matrices for matrix multiplication $H = F \cdot G$.

In spite of the similarity to both previous algorithms there are two essential differences:

a) a partial matrix and not a single element never mind a bit is assigned to each processor,

b) each processor has its own control and starts multiplying its partial matrices at the moment the data arrived at its memory (local instead of global synchronization).

The first point was the reason for naming this algorithm "macro systolic". Both features of distinction demonstrate two advantages of an EGPA-system in comparison to a systolic array:

a) The number of processors does not restrict the magnitude of the matrices. The size of a partial matrix is only bounded by the memory attached to each processor.

b) Differences in the time for computing the partial products will be equalized. In the case of a systolic array the worst case always determines the duration of one step.

At the end of this paragraph we give speedups measured for dense matrices at the pilot pyramid:

Size n of n x n matrix	Speedup
8	1
100	3,56
200	3,70

The factors decrease with decreasing n. The reasons are the approximately constant effort for the setup and the increasing portion for synchronization.

A theoretical speedup for systems of any size is given in [1o] (speedups near the number of processors in the working plane, if the size of the matrices is big enough). In the mentioned paper there is listed an improvement of the algorithm, too.

4. AN 85-PROCESSOR-SYSTEM

The EGPA-architecture allows for the extension towards more powerful multiprocessor systems by adding one or more larger processor arrays at the bottom of the pyramid. At the Computer Science Department (IMMD) of the University of Erlangen-Nürnberg the Erlangen Multiprocessor System 85 (EMSY 85) has been conceived on the base of the experience gained with the EPGA-project. Four arrays of 1, 4, 16 and 64 processor-memory-modules (PMMs) can respectively be hierarchically arranged. The elementary structure is a pyramid equivalent to the EGPA-pilot-pyramid. For example each PMM may consist of an iAPX 286/287 microprocessor and a one-half-megabyte multiport memory. The operating system may be based on UNIX and can analogously be organized to the hardware structure. The operating system could be distributed over the 85-pyramid, increasing in power towards the top [2]. The structural concept of the 85-pyramid is depicted in Fig. 14.

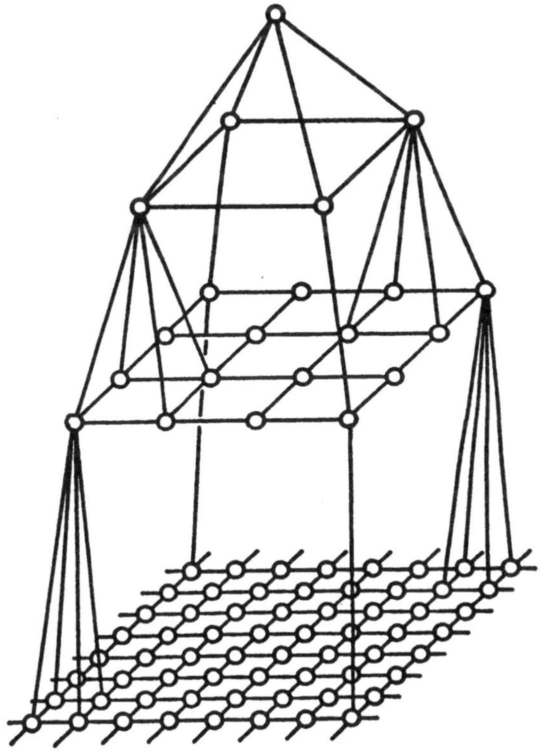

Fig. 14: An extended EGPA-system consisting of four processor planes with 1, 4, 16 and 64 processors.

Such an extended EGPA-system would especially be appropriate for multigrid applications. Presently, studies are being carried out with regard to the infrastructure of the PMMs, details of the connection system , I/O-performance etc. The research work is aimed at a multiprocessor system - based on the EGPA conception - which allows the efficient computation of a wide class of multigrid problems.

5. REFERENCES

[1] Cannon: Ph.D. Thesis. Montana State University 1969

[2] Fritsch, G.; Kleinöder, W.; Linster, C.-U.; Volkert, J.: EMSY 85 - The Erlangen multiprocessor system for a broad spectrum of applications. Proc. 1983 Int. Conf. Parallel Processing, IEEE Comp. Soc. Order No. 479 (ed. H.J. Siegel and L. Siegel), 325-330, IEEE Computer Society Press 1983.

[3] Fritsch, G.; Müller, H.: Parallelization of a minimisation problem for multiprocessor systems. Lect. Notes in Computer Science, No. 111 (ed. W. Händler), 453-463, Springer Verlag 1981

[4] Fromm, H.J.: Multiprozessor-Rechneranlagen: Programmstrukturen, Maschinenstrukturen und Zuordnungsprobleme. Arbeitsberichte des IMMD, Universität Erlangen-Nürnberg, Band 15, Nr. 5, 1982.

[5] Geus, L.; Henning, W.; Vajtersic, M.; Volkert, J.: Matrix Inversion Algorithm for the parallel Computer EGPA. Interner Bericht des IMMD der Universität Erlangen-Nürnberg, Veröffentlichung inVVorbereitung

[6] Goessmann, M.; Volkert, J.; Zischler, H.: Image Processing and Graphics on EGPA. EGPA-Bericht 1982, IMMD, Universität Erlangen-Nürnberg

[7] Händler, W.; Herzog, U.; Hofmann, F.; Schneider H.J.: Multiprozessoren für breite Anwendungsbereiche: Erlangen General Purpose Array. Architektur und Betrieb von Rechensystemen. Informatik Fachberichte Nr. 78, Berlin 1984

[8] Händler, W.; Hofmann, F.; Schneider, H.J.: A General Purpose Array with a Broad Spectrum of Applications, Vol. 4, pp. 311-335 in: Informatik Fachberichte, ed. Händler,W., Springer-Verlag, Berlin-Heidelberg-New York, 1976

[9] Händler, W.; Rohrer, H.: Gedanken zu einem Rechner-Baukasten-System. Elektronische Rechenanlagen, Vol. 22, Nr. 1, pp. 3-13, 1980

[10] Henning, W.; Volkert, J.: Multiplikation von Matrizen auf EGPA-Multiprozessoren. PARS Workshop, Erlangen, 1984

[11] Hoffmann, U.; Panthel, G.; Emig G.: FIBSAS Standardprogramm zur Auslegung und Analyse von Schüttschichtreaktoren. Bericht des Instituts für Technische Chemie, Universität Erlangen-Nürnberg, 1977

[12] Katona, E.: Cellular algorithms for binary matrix operations. in: Cellprocessors and Cellgorithms, Report on a common research project, Nr. 1, p. 129, Technische Universität Braunschweig, Informatik-Skripten, ed. Vollmar, R., and Legendi, T.

[13] Kung, H.T., Leiserson, C.E., Systolic Arrays (for VLSI). Spare Matrix Proc., Duff, et al., eds., Society of Indust. and Appl. Math.

Philadelphia, Pa., pp. 245-282, 1978

[14] Legendi, T.: Cellprocessors in computer architecture. in: Cellprocessors and Cellgorithms, Report on a common research project, Nr. 1, p. 1, Technische Universität Braunschweig, Informatik-Skripten ed.: Vollmar, R.; and Legendi, T.

[15] Maehle, E.: Modulare, fehlertolerante Multimikroprozessorsysteme nach dem Baukastenprinzip. VDI-Berichte, 395, 91-96, 1981

[16] Nelder, J.A., Mead, R.: A simplex Method for Function Minimization. Comp. J., Vol. 7, 308-313, 1965

[17] Rathke, M.: SAP - Ein optimistischer Algorithmus für die parallele Textverarbeitung. EGPA - Internal Paper (to be published)

[18] Seidl, W.: Berechnung der Minima der Zielfunktion aus dem Programmsystem FIBSAS auf dem EGPA-Mehrrechner-System. Diplomarbeit, IMMD, Universtität Erlangen-Nürnberg, Erlangen 1984

[19] Viegas de Vasconcelos, A.; Hosemann, G.: Transient Studies on a Multiprocessor. IEEE-Power Engineering Society /Winter Meeting 1984 (Dallas), Paper No. 84 WM 096-4. To be published in IEEE-Transactions on Power Apparatus and Systems

FLEXIBLE PROCESSOR ARRAY WITH PARALLEL MEMORY
(Foundations of the basic architecture)

M. Gössel and B. Rebel[1)]

Introduction

At our institute we have designed a parallel computer. The main components of this computer are the quadratic flexible processor array and the parallel memories with recursive address generation.

The basic structure of the system is shown in Fig. 1.

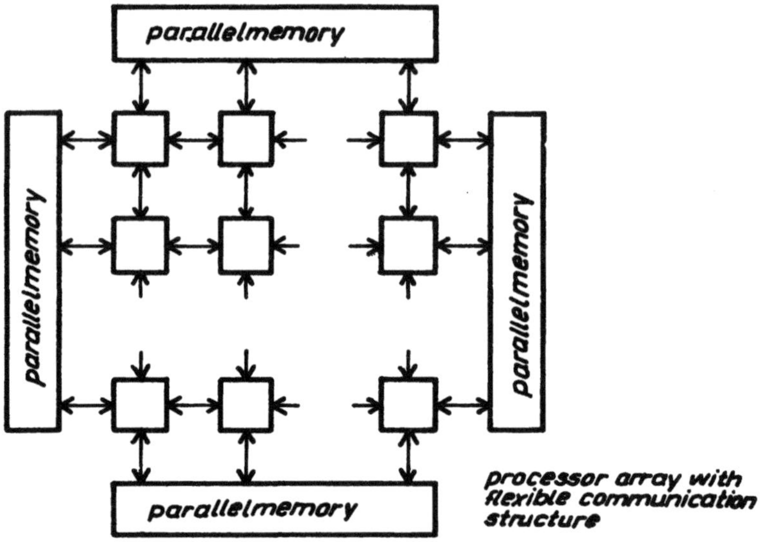

Fig. 1

The architecture of our system is influenced by picture processing problems. A quadratic processor array is supplemented with four parallel memories on its sides.

In this paper we would like to present some basic theoretical considerations which the architecture of our system is based on.

The topics we want to discuss here are:

---

1) Zentralinstitut für Kybernetik und Informationsprozesse der Akademie der Wissenschaften der DDR, DDR-1086 Berlin, Kurstraße 33

1. The influence of the choices of the memory function and address function on the architecture of the memory. These functions determine the possible access formats or memory windows, the possibilities of decentralized address generation and of recursive address generation.

2. The interrelation between memory windows and operator windows, especially the problem of access optimality of gliding operators and its implications.

3. Why do we use a processor array with flexible communication structure and with a bidirectional register net.

In /11/ the structure of our system is explained in greater detail and it is shown how to implement parallel algorithms.

Basic notions and notations

Fig. 2 shows a digital image of a horse in a scanning field R and a parallel memory. A parallel memory consists of N memory modules $S_0, \ldots, S_{N-1}$, an address generator A and a permutation network π .

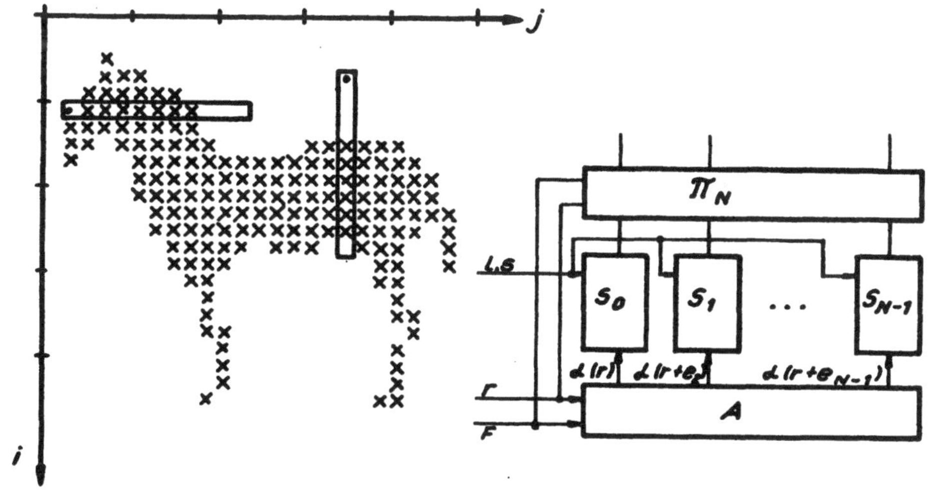

Fig. 2

If a digital image is stored in a parallel memory, for every pixel r of the scanning field R the memory module and the address within the memory module where the value v(r) at pixel r is stored is to be determined.

Formally two functions are to be defined, the module assignment function S(r) and the address function $\alpha(r)$. The module assignment function $S: R \longrightarrow \{0,\ldots,N-1\}$ assigns the index S(r) of the memory module, where the value v(r) is stored to every pixel r of R.

The address function $\alpha: R \longrightarrow \{0,\ldots, a(max)\}$ determines the address a(r) under which v(r) is stored in S(r).

To compute a local operator O the processor needs a set of data in a special data format or an operator window F_o.

Formally an operator window F_o is a rigid template which can be described by a set of vectors

$$F_o = \{v_1, v_2, \ldots, v_K\} \quad v_1 = (0,0,\ldots,0), \; (v_i = v_i^1, \ldots, v_i^m)$$

where v_i^j are integers, $m > 2$.

A window $F_o(r)$ at the point $r \in R$ is the subset

$$F_o(r) = \{r, r+v_2, \ldots, r+v_K\}$$

of the scanning field R, where r is called the emphasized point of the window.

Formally a memory window F_s is also a rigid template which can be described by a set of vectors

$$F_s = \{e_1 e_2, \ldots, e_M\}, \; e_1 = (0,\ldots,0), \; e_i = (e_i^1, \ldots, e_i^m)$$

where e_i^j are integers, and $F_s(r)$ is the subset

$$F_s(r) = \{r, r+e_2, \ldots, r+e_M\}$$

of the scanning field R.

A memory window F_s describes a set of data which can be accessed in parallel from a parallel memory. Therefore it is necessary to store all values of a memory window in different memory modules. If this is the case for a given window F and a module assignment function S, then F is called F-conflictfree.

<u>Definition 1.</u> A module assignment function S is called conflictfree with respect to a window F, if we have for $r \in R$, $r', r'' \in F(r)$ and if $r' \neq r''$ implies $S(r') \neq S(r'')$.

<u>Definition 2.</u> A module assigment function S is called conflictfree with respect to a set of windows $\mathbf{F} = \{F_1, \ldots, F_L\}$ if S is F_i-conflictfree for every F_i of $\mathbf{F}$.

Fig. 3 shows a memory function $S(i,j) = (2i+j) \bmod 5$ and conflictfree windows F_1, F_2, F_3 and F_4.

Windows of special interest are generalized straight lines and parallelograms of generalized straight lines.

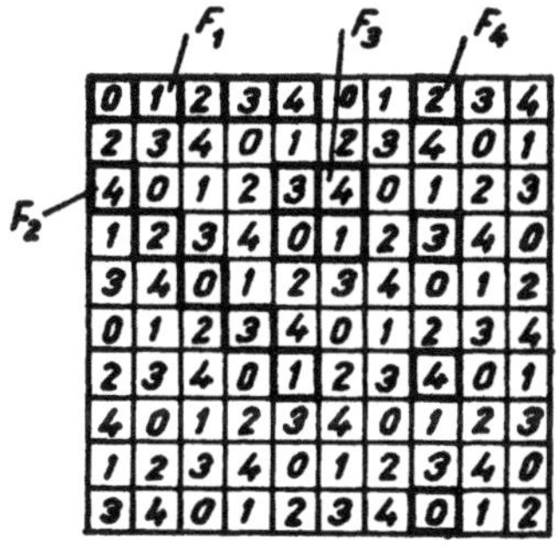

Fig. 3

A window $G(r) = \{r+e_1,\ r+e_2, \dots, r+e_M\}$ is a generalized straight line if $e_k = (k-1)e_2$, $k = 1,\dots,M$. The vector e_2 is the defining vector of G.

A parallelogram P(r) of generalized straight lines G is the set

$$P(r) = \bigcup_{l=1}^{L} G(r+(l-1)\tilde{e}_2) = \bigcup_{m=1}^{M} \tilde{G}(r+(m-1)e_2)$$

with

$$G(r) = \{r'/r' = r+(k-1)e_2,\ k = 1,\dots,M\}$$

and

$$G(r) = \{r''/r'' = r+(n-1)\tilde{e}_2,\ n = 1,\dots,L\} .$$

Various generalized straight lines and parallelograms of straight lines are represented in Fig. 4.

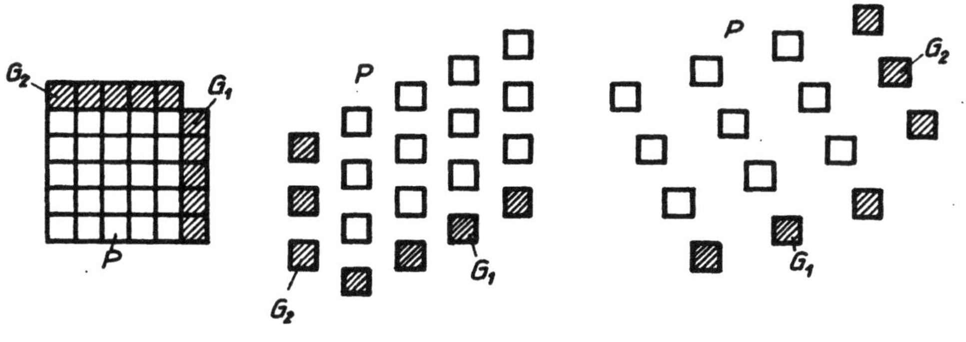

Fig. 4

Memory and address functions and their influence to the memory architecture

Different classes of memory functions such as periodic functions, isotropic function (called regular functions in /15/) and linear functions are investigated in /1, 3, 7, 9, 10, 12, 14, 15/.

It is well known that every linear function is isotropic, that every isotropic function is periodic and that every isotropic function is isomorphic to a linear function it the number of memory modules is a prime number /9, 10, 12, 15/.
Linearity means here linearity with respect to the module of a natural number. The module assignment function of /1, 2/ is linear in $GF(2^n)$ but not isotropic. Since the early paper /12/ the close interrelation between tesselation of a plane by a window F and the existence of an F-conflictfree memory function has been clear. A special chapter of /12/ is devoted to conflictfree access if the windows are straight lines and generalized straight lines. As a special result module assignment function modulo a prime number is F_G-conflictfree where F_G is the set of almost all generalized straight lines.

Now we show that for arbitrary conflictfree windows the address computation can be decentralized if the memory function is isotropic /9/.

First we introduce the notion of an isotropic module assignment dunction.

Definition 3. A module assignment function $S: R \longrightarrow \{0,1,\ldots,N-1\}$ is called isotropic if there exist uniquely invertible functions

$$\emptyset_x: \{0,1,\ldots,N-1\} \longrightarrow \{0,1,\ldots,N-1\}$$

$$\emptyset_y: \{0,1,\ldots,N-1\} \longrightarrow \{0,1,\ldots,N-1\}$$

$$\emptyset_z: \{0,1,\ldots,N-1\} \longrightarrow \{0,1,\ldots,N-1\}$$

such that we have for all $r = (i,j,k) \in R$

$$S(i+1,j,k) = \emptyset_x(S(i,j,k)),\ S(i,j+1,k) = \emptyset_y(S(i,j,k))$$

$$S(i,j,k+1) = \emptyset_z(S(i,j,k)),$$

where the functions $\emptyset_x$, $\emptyset_y$, $\emptyset_z$ are mutually commutative, i.e.

$$\emptyset_x\emptyset_y = \emptyset_y\emptyset_x,\ \emptyset_x\emptyset_z = \emptyset_z\emptyset_x,\ \emptyset_y\emptyset_z = \emptyset_z\emptyset_y.$$

Now we prove

Theorem 1. Let $F = \{e_1, e_2, \ldots, e_M\}$ be a window and $S: R \longrightarrow \{0,\ldots,N-1\}$ an isotropic F-conflictfree module assignment function. Then there exists such a function $\chi: \{0,\ldots,N-1\} \longrightarrow F$, that we have for all $r \in R$ and $P \in \{0,\ldots,N-1\}$

$$S(r+\chi(S(r),P)) = P \qquad (x)$$

Proof. We put

$$\chi(S,P) = e_j \text{ iff } \phi_x^{e_1^j} \phi_y^{e_2^j} \phi_z^{e_3^j} (S) = P .$$

Since

$$S(r+e_j) = \phi_x^{e_1^j} \phi_y^{e_2^j} \phi_z^{e_3^j} (S(r))$$

eq. (*) is fulfilled.

For the linear case we have

$$S(i,j,k) = a.i+b.j+c.k \pmod N$$

which implies

$$\phi_x (S(i,j,k)) = S(i+1,j,k) = (S(i,j,k)+a) \mod N,$$

$$\phi_y (S(i,j,k)) = S(i,j+1,k) = (S(i,j,k)+b) \mod N,$$

$$\phi_z (S(i,j,k)) = S(i,j,k+1) = (S(i,j,k)+c) \mod N,$$

$$\chi(P,S) = e_j \text{ iff } \phi_x^{e_1^j} \phi_y^{e_2^j} \phi_z^{e_3^j} (S) =$$

$$= (s+e_1^j a+e_2^j.b+e_3^j c) \mod N = P$$

and finanlly

$$\chi(S,P) = e_j \text{ iff } (P-S = e_1^j a+e_2^j.b+e_2^j c) \mod N,$$

and χ depends on P-S only.

Jùst for a moment let us consider the module assignment function $S(i,j) = (i \mod 2^n) \oplus (j \mod 2^n)$, where $(i \mod 2^n)$ and $(j \mod 2^n)$ are represented as binary numbers and the operation "$\oplus$" is a componentwise modulo 2 addition /1, 2/. For $n = 3$ $S(i,j)$ is represented in Table 1.

		0	1	2	3	4	5	6	7
		000	001	010	011	100	101	110	111
0	000	0	1	2	3	4	5	6	7
1	001	1	0	3	2	5	4	7	6
2	010	2	3	0	1	6	7	4	5
3	011	3	2	1	0	7	6	5	4
4	100	4	5	6	7	0	1	2	3
5	101	5	4	7	6	1	0	3	2
6	110	6	7	4	5	2	3	0	1
7	111	7	6	5	4	3	2	1	0

The module assignment function is not isotropic although it is linear in $GF(2^n)$. This function does not allow a decentralized address computation.

Now we discuss the possibility of a recursive decentralized address computation. A recursive, decentralized address computation is especially well adapted to problems of image processing when an operator with an window F is gliding step by step over the picture. A recursive decentralized address computation is possible if both the module assignment function and the address function are isotropic.

We suppose $S(i,j,k)$ to be an isotropic module assignment function as defined in Definition 3 and $\alpha : R \longrightarrow \{o,\dots,a(max)\}$ to be an isotropic address function with

$$\alpha(i+1,j,k) = \psi_x(\alpha(i,j,k)), \alpha(i,j+1,k) = \psi_y(\alpha(i,j,k)),$$

$$\alpha(i,j,k+1) = \psi_z(\alpha(i,j,k)).$$

The two successive positions of the considered window $F = \{e_1,\dots,e_M\}$ are $r = (i,j,k)$ and $r+dr$ with $dr = (di, dj, dk)$ where di, dj, dk are the relative increments of i, j, k.

The decentralized recursive address computation can be realized in the following steps:

The address computer A_p, $P \in \{0,\dots,N-1\}$ computes

1. $S(r+dr) = \phi_x^{di}\,\phi_y^{dj}\,\phi_z^{dk}\,(S(r))$, the memory module of the emphasized scanning point of F in its new position,
2. $\chi(S(r+dr),P) = (u_1, u_2, u_3)$, the distance between the emphasized scanning point in new position and the pixel of F stored in memory module P,
3. $\alpha\,(r+dr+\chi(S(R+dr), P))) = \psi_x^{di+u_1}\,\psi_y^{dj+u_2}\,\psi_z^{dk+u_3}\,(\alpha(r))$, the address of the pixel of F (r+dr) stored in memory module P.

A parallel memory with recursive, decentralized address computation is represented in Fig. 6.

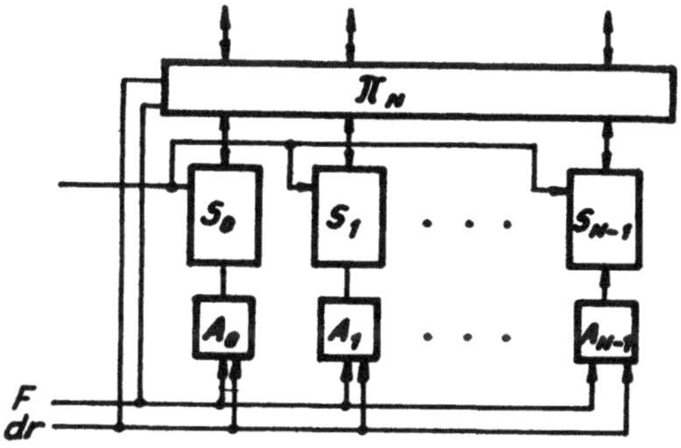

Fig. 5 Fig. 6

If both the module assignment function and the address function are linear these computations are quite simple. For example Table 2 shows a 2-dimensional scanning field with $S(i,j) = (2i+j) \bmod 5$ and $\alpha(i,j) = (10i+j) \bmod 20$.

0	1	2	3	4	0	1	2	3	4
0	1	2	3	4	5	6	7	8	9
2	3	4	0	1	2	3	4	0	1
10	11	12	13	14	15	16	17	18	19
4	0	1	2	3	4	0	1	2	3
0	1	2	3	4	5	6	7	8	9
1	2	3	4	0	1	2	3	4	0
10	11	12	13	14	15	16	17	18	19

Table 2

A first effort of a recursive address computation was published in /8/, but the recursive address computation was not realized with the addresses itself but with an auxilary function $A(i,j) = i+j$, much greater than the addresses itself and therefore unnecessarily expensive.
Now we discuss the interrelation of memory windows and operator windows for gliding operators and restrict ourselves to 2 dimensions only.

To compute a local operator $O(r)$ at the point r of R the data located in the operator window $F_o(r)$ are needed.
Since the picture is supposed to be stored in a parallel memory, $F_o(r)$ has to be composed of several subsequentially accessed memory modules and $F_o(r)$ has to be a subset of the set-theoretical union of K memory windows $F^i_s(r_i)$,

$$F_o(r) \subseteq F_s^1(r_1) \cup F_s^1(r_2) \cup \ldots \cup F_s^K(r_K) .$$

Since $e_1 = (0)$ and $F_o(r) = r, r+v_2,\dots,r+v_M$

we have

$$F_o(r) \subseteq F_s^1(r) \cup F_s^1(r+v_2) \cup\dots\cup F_s^1(r+v_M),$$

and hence K = M.

Now we suppose that the local operators $0_1,\dots,0_2$ are subsequently applied to different points r_i of R. For every corresponding operator window we need K_i memory accesses. $1 \leq K_i \leq M_i$.

Let Z be the number of memory accesses for computing $0_1(r_1),0_2(r_2),\dots, 0_L(r_L)$.

Then we have $L \leq Z \leq \sum_{i=1}^{L} K_i$, and for $K_1=K_2=\dots=K$

$$L \leq Z \leq K.L .$$

For overlapping successive operator windows the number of memory accesses can be reduced. Computing a sequence of operators $0_1,\dots,0_2$, the smallest number of memory accesses we can obtain is

$$Z = L + c , \quad c = \text{const.} \qquad \text{(xx)}$$

A sequence of operators for which eq. (xx) is valid is called memory access optimal.

Now let uns consider an oprator 0 gliding in constant stepts dr over the scanning field.

We suppose 0 to be memory access optimal. Therefore, after computing 0(r) only one additional memory access is necessary to compute 0(r+dr) and we conclude that

$$F_o(r+dr) = F_s(r+dr) \cup (F_o(r) \setminus H(r))$$

with the reasonable conditions

$$H(r) \subseteq F_o(r)$$

and

$$F_s(r+dr) \cap F_o(r) = \emptyset$$

is valid.

If we suppose additionally that 0 is able to glide access optimally with respect to dr in steps dr, dr $\neq$ a.dr, a = const, then we have

$$F_o(r+d\tilde{r}) = \tilde{F}_s(r+d\tilde{r}) \cup (F_o(r) \setminus \tilde{H}(r)),$$

$$\tilde{H}(r) \subseteq F_o(r) ,$$

$$\tilde{F}_s(r+d\tilde{r}) \cap F_o(r) = \emptyset .$$

Then it is easy to proof the following

Theorem. Let O be a local oprator with the operator window $F_o(r)$ and let

$$F_o(r) = \bigcup_{l=1}^{N} G(r+(l-1)e_2)$$

with

$$G(r) = r'/r' = r+(k-1)\tilde{e}_2, k=1,\dots,N$$

be a parallelogram of generalized straight lines of N^2 pixels, then O is access optimal for dr = $\pm e_2$ with the memory window

$$F_s = \{e_1, e_2, \dots, e_N\} \text{ with } e_l = (l-1)\tilde{e}_2,\ l=1,\dots,N$$

and for $d\tilde{r} = \pm \tilde{e}_2$ with the memory window

$$F_s = (e_1^* \ e_2^*, \dots, e_N^*) \quad \text{with } e_l^* = (l-1)e_2, l=1,\dots,N.$$

Fig. 4 shows different parallelograms of straight lines and the corresponding memory window for access optimal gliding operator windows.

Furthermore it can be shown that a gliding operator is access optimal if and only if its operator window is a subset of parallelograms of generalized straight lines with distances between adjacent scanning points of dr and $d\tilde{r}$ respectively.
Due to this fact we use parallel memories with conflictfree parallel access to generalized straight lines as described in /7, 9/ and a processor array with a structure as shown in Fig. 1.

To obtain as many as possible conflictfree windows in the form of a generalized straight line the module assignment function is chosen to be $S(i,j,k) = (a.i+j+b.k) \bmod N$ where N is a prime number.

The different scanning points of a generalized straight line $G(r) = \{r, r+e_2, r+2e_2, \dots, r+(N-1)e_2\}$ are stored in memory modules $S(r), (S(r)+\delta) \bmod N, \dots, (S(r)+(N-1)\delta) \bmod N$ with $\delta = (a.e_2^1+e_2^2+b.e_2^3) \bmod N$.

Since N is prime, for $\delta \neq 0$ we have
$\{S(r), (S(r)+\delta) \bmod N, \dots, (s(r)+(N-1)\delta) \bmod N\} = \{0,1,\dots,N-1\}$,
and for every generalized straight line with $\delta \neq 0$ S is conflictfree.

If an operator with an operator window that is a parallelogram of generalized straight lines is to be computed the processor has to know the positions of the grey-values of the pixels in the array. After gliding a step dr resp. $d\tilde{r}$ these positions have changed by dr and dr respectively.

These changes can be realized by means of a two dimensional register structure with bidirectional communication lines as implemented in our array.
The flexibility of the array is acieved by the possibility to put the

connections through the registers.

This possibility allows us to adapt the processor array to different algorithms as illustratet for a tree in Fig. 7.

Fig. 7

This approach needs less pins than the flexible structure described in /13/.

Moreover the system is fault-tolerant. If a processor fails, the corresponding row and column of the array are put through and the size of the NxN array is diminished to (N-1)x(N-1).

Using the described flexibility different systems can be simulated.

Sumarizing, our system is especially well adapted to

- gliding operators in picture processing,
- systolic algorithms,
- a structural embedding of algorithms into the array

and therefore it is relatively simple to program the proposed processor array.

References

/1/ Batcher, K. E.: The multidimensional access memory in STARAN. IEEE Trans. Comp. C-26, 174-177, (1977).

/2/ Batcher, K. E.: Multidimensional access solid state memory. US PS 3800289.

/3/ Budnik, P., Kuck, D. J.: The organization and use of parallel memories. IEEE Trans. Comp. C-20, 1566-1571 (1971).

/4/ Gössel, M., Rebel, B.: Parallel memory with recursive address computation. Proc. Parallel Computing 83, (ed.) Feilmeier, M,... Elsevier Sc. Publ. 1984.

/5/ Gössel, M., Rebel, B.: Gliding operators in image processing and memory access for parallel computers. Beiträge zur Theorie der Polyautomaten, Informatik-Skripten 8, 29-40, 1984. Herausg. R. Vollmar und U. Golze, Univ. Braunschweig.

/6/ Graf, A., Pöschel, R.: Special Communications of processors in an array for parallel computing, presented at II. Int. Workshop, Parallel processing by cellular automata and arrays, Berlin 1984.

/7/ Lawrie, D. H.: Access and alignment of data in an array processor. IEEE Trans. Comp. C-24, 1148-1156 (1975).

/8/ Lawrie, D. H., Vora, R.: The prime memorie system for array access. IEEE Trans. Comp. C-31, 435-442 (1982) (s. USPS 4051551).

/9/ Rebel, B., Gössel, M.: Ein paralleler Speicher, Forschungsbericht ZKI, Berlin, 1982 (s. DDR PS 208499).

/10/ Rebel, B., Gössel, M.: On conflictfree parallel memory access. Computers and Artificial Intelligence 2, 395-401 (1983).

/11/ Rebel, B., Wilhelmi, W.: Specialarchitecture features of a new image processor system presented at II. Int. Workshop, Parallel processing by cellular automata and arrays, Berlin, 1984.

/12/ Shapiro, D. H.: Theoretical limitations on the use of parallel memories, Ph. D. Thesis, Dep. Comp. Sc., Univ. Illinois at Urbana Champain, UIUCDS-R-75-776 (1975).

/13/ Snyder, L.: Introduction to the configurable, highly parallel computer, Computer 15, 47-56 (1982).

/14/ Voorhis, D. C., Morrin, T. H.: Memory system for image processing. IEEE Trans. Comp. C-27, 113-125 (1978) (s. DEPS 2747075).

/15/ Wijshoff, H. A. G., v. Leeuven, J.: The structure of periodic storage schemes for parallel memories, Techn. Rep. RUU-CS-84-1 Dep. Comp. Sc., Univers. Utrecht 1984.

SORTING ON A BIT ORGANISED PARALLEL COMPUTER

D.Parkinson and C. Boyle

DAP Support Unit

Queen Mary College

Mile End Road

London E1 4NS

England

Introduction

Sorting has always been an important operation in computer applications and it is therefore natural to search for sorting strategies for parallel computers. There have been many studies of sorting algorithms for various parallel computers (1-7). All papers on parallel algorithms have to assume some model of parallel computer and the differences between many of the papers arise directly from differences in the facilities of the model computers.

Important criteria include:

- What is the connectivity network assumed between the processors - (mesh connected, perfect shuffle etc.)?
- Do the processors possess local indexing capability?
- What is the ratio between the number of items and the number of processors?

The prime model of processor in all studies appears to have been a processor which supports a basic unit of computing "The comparison/exchange" and all studies seek to minimise the number of comparison exchanges and/or the number of routing operations. Array processors with many thousands of processors have been built (DAP-4096 processors, CLIP-9216 processors and MPP-16384 processors)(8). The basic processors in these systems are bit organised and some of these systems have the capability of supporting associative or content addressable algorithms.

The objective of this paper is to examine sorting algorithms which exploit the associative nature of bit organised array processors.

The Basic Computational Model

The ICL-DAP is the prime model of processor array which we will consider. The features that are relevant to our study are shown in figure 1. There are N processors, each with private memory under the control of a master processor in an SIMD fashion. Processors may communicate either locally, via neighbourhood connections, or via a highways to the master processor. In the case of the DAP the N(=4096) processors are connected in a 2-dimensional mesh and there are two sets of highways (rows and columns) onto which each processor is capable of "anding" a bit. The operation takes only one machine cycle and so global testing for the presence of a single bit (B) anywhere in the array is performed by the following set of operations:-

Simultaneously AND(NOT.B) onto the highway and put the result into a register R in the Master Processor.

Test R - if R is false there is at least one processor with its local value of B with value 'true'.

FIGURE 1.

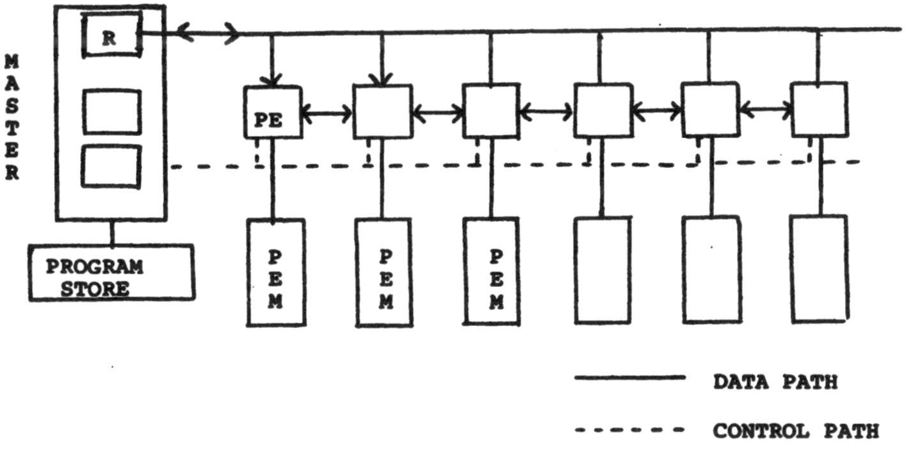

It is essential to the model of computation that both these operations take place only one machine cycle each and not $\log_2 N$ cycles needed in other models of connectivity. The fact that the DAP does have these facilities shows that the model is not unreasonable. This model of computation does not assume that the system has the ability (in one cycle) to count how many processors have the bit B set, such a facility would be helpful for speeding up parts of many applications. On the other hand, a routine to count the number of bits set can exploit the bit organised nature of the processors and match the precision of intermediate arithmetic with the maximum known precision. On the DAP this COUNT routine takes approximately 100 machine cycles to count the maximum of 4096 element - this compares very favourably with about 50 machines cycles needed for general 12 bit precision fixed point arithmetic within a single processor.

Given these facilities the largest element in the array may be selected by a well know associative algorithm (9). Suppose each processor has a p bit field and b_i is the value of the i-th bit with b being the most significant bit. Th algorithm may be expressed as follows:

```
candidates: = true ; mark all items as possible maxima
for i = 1 to p begin.
        if any (candidates.and.b(i)) then
           candidates = candidates.and. b(i)
        end.
```

This algorithm takes a time proportional to the p - the length of the field and is independent of the number of processors. A simple extension of the above algorithm allows the maximum element to be identified over any subset of the processors. Using the algorithm it is possible to create a 'sort by selection' algorithm on our model computer with an execution time O(N).

The existence of a fast maximum (or minimum) algorithm does lead to the question "Why do we wish to sort on processors with associative searching?". This is an important question and can only be properly answered when the context is taken into account. Often sorting is a means of obtaining efficient serial access to data and is in conflict with the assumption of parallelism. In practise, we might wish to sort as part of a larger computation where all the data cannot be held with the system memory, i.e. as an "in-core" sort subset of a larger "out-of-core" sort. Another reason for sorting might be as a means to answer questions of the type "Is any data record duplicated?" One way of answering this last question is to sort the data and then compare adjacent items.

Alternative Sorting Algorithms

The selection sort discussed above is not based upon the repeated application of comparison-exchange operations and it is natural to ask if there are other sorting algorithms which can exploit the special properties of our computational model. As the 'QUICKSORT" class of algorithms are usually to be the most efficient serial algorithms, it is reasonable to study them as starting points for a parallel implementation. There are many ways of viewing the QUICKSORT algorithm but the best from our point of view is the following:

QUICKSORT - basic step for a M element set.

Step 1 Select a member of the set call it 'P'.

Step 2 Compare all members of the set with P marking them as being either less than or greater than P.

Step 3 Separate the set into two sets, all members of one set being less than P and all members of the other set being greater than P.

Step 4 Place the element P between the two sets produced in step 3.

The above basic step reduces the initial set into two sets and the algorithm proceeds by repeating the basic step on the smaller sets. The usual description (e.g. ref. 1) of the algorithm do not describe the method in the way presented above, but perform Step 3 and 4 in a integrated fashion by means of pairs of pointers. Some variations of the QUICKSORT algorithm only differ in the strategy used to select the element P in Step 1. The optimal strategy is one which would always divide the initial set into two equal smaller sets.

If we adopt the approach given above we can see that Step 2 is ideally suited to the parallelism of our model computer. Step 1 loads the test value into the master processor. Step 2 simultaneously compares the local content of a processor with the test value and creates a true or false result. Step 3 requires that all items in the true class are collected together as are all those in the false class. Step 4 then requires that the comparison value is placed in a correct location.

When we try to implement the algorithm on our bit organised processor array we first want to derive a strategy for selecting the comparison element. The desire is to choose a value which has a high probability of splitting the data set in two equal parts. Figure 2 shows schematically a set of data values on a 8 PE machine. We see that steps 1 and 2 may be combined in zero time if we select the most significant bit in the key field as an indicator which divides the data set into two groups (step 4 also vanishes).

Figure 2

PE	1	2	3	4	5	6	7	8		
	1	0	1	1	0	1	0	1	b_2	
	0	1	0	0	1	1	0	1	b_1	KEY
	1	0	1	0	1	0	1	1	b_0	
	(5)	(2)	(5)	(4)	(3)	(6)	(1)	(7)		value of KEY

If the Keys are uniformly distributed amongst the possible values in the range we can expect there to be (on average) an equal number of 0's and 1's at any bit location. If we adopt the convention that we choose the most significant bit position as a discriminator, we then have performed (for nothing) all the 'comparisons' implied in the algorithm. We are only left with the task of implementing Step 3 which requires us to compress data left or right depending on the value of the bit in location b i.e. we would like to transform the data to the state shown in Figure 3. The method by which such compressions may be made is dependant on the details of the system. A general software compression technique is described in the Appendix.

Figure 3

PE	1	2	3	4	5	¦	6	7	8
	1	1	1	1	1	¦	0	0	0
	0	0	0	1	1	¦	1	1	0
	1	1	0	0	1	¦	0	1	1
	(5)	(5)	(4)	(6)	(7)	¦	(2)	(3)	(1)

The data is now divided into two sets and a straightforward application of the principle would examine location b which would identify the data to be compressed left and right. There is an additional complication as these compressions are to be done only within the groups previously established, i.e. in the example no data is to move across the boundary between Processors 5 and 6.

On the assumption that we can effect this transformation the new location of our sample data would be come:

Figure 4

PE	1	2	3	4	5	6	7	8
	1	1	1	1	1	0	0	0
	1	1	0	0	0	1	1	0
	0	1	1	1	0	0	1	1
	(6)	(7)	(5)	(5)	(4)	(2)	(3)	(1)

The final operation using b_0 but preventing compressions across boundaries 2/3, 5/6 and 7/8 would lead to fully sorted data in Figure 5:

Figure 5

PE	1	2	3	4	5	6	7	8
	1	1	1	1	1	0	0	0
	1	1	0	0	0	1	1	0
	1	0	1	1	0	1	0	1
	(7)	(6)	(5)	(5)	(4)	(3)	(2)	(1)

We call this algorithm TOP-DOWN Compression (TDC).

In order to implement this algorithm we would need to have a procedure for performing the appropriate compressions left (and right). As we have described the algorithm the procedure(s) would have to be capable of limiting their operations within areas defined by a continuously up-dated series of boundaries. This last requirement can be eliminated if we make use of an observation which dates from the days when mechanical card sorters were used, that is, to perform the selections (compressions) starting at the least significant bit. This algorithm is the Bottom-Up Compression (BUC) algorithm.

If the least significant bit was used on the first step, Figure 3 would become:

Figure 3b

PE	1	2	3	4	5	6	7	8
	1	1	0	0	1	0	1	1
	0	0	1	0	1	1	0	1
	1	1	1	1	1	0	0	0
	(5)	(5)	(3)	(1)	(7)	(2)	(4)	(6)

compressing using b_1 across all the processors would now give the data shown in Figure 4b:

Figure 4b

PE	1	2	3	4	5	6	7	8
	0	1	0	1	1	1	0	1
	1	1	1	1	0	0	0	0
	1	1	0	0	1	1	1	0
	(3)	(7)	(2)	(6)	(5)	(5)	(1)	(4)

and a final stage of compression using b_2 would result in the fully sorted data as in Figure 5.

This modification means that we only need a procedure for compressing the entire array according to a prescribed mask. The cost we have to pay for reducing the complication in the compression routine is the possibility that data is routed totally across the array every step, whereas the alternative approach has shorter and shorter routing operations.

Summary

This work has been sufficient to show that the Quicksort type of algorithms can be efficiently implemented on a bit organised processor array. The execution time of the algorithms will depend on the range of key values as well as the detailed facilities of the hardware. In a future paper we hope to conduct a more detailed comparison of the complexity of the algorithms presented here and the bitonic sort algorithms.

APPENDIX

Compression Operations

The approach we are investigating has replaced the usual comparison exchange operations with a number of compression operations. The relative efficiency of these compression based algorithms compared with other algorithms will depend crucially on how well any particular processor array can implement the operations. One option would be to build such operations into the hardware of the processor array as is sometimes done in vector processors.

The technique used on the DAP to perform the data compression operations proceeds in two phases:

Phase 1 - computation of distance each item is to move.

Phase 2 - movement of data items in blocks.

The input to the operation includes a bit map which is set to the value 0 in those processsors which contain data to be retained and has the value 1 in those processors where the data is to be rejected. Phase 1 then is accomplished by creating in each processor the sum of all bits in the processors to the left (if compression left is needed). Such a sum is obtained using the well known recursive doubling algorithm modified to exploit the very low precision of numbers in the early steps and so has low arithmetic complexity.

Phase 1 leaves in each processor a number of 'n' which indicates how many places left must the contents be moved. The actual movement is implemented in Phase 2 by examining the bit patterns which make up the numbers which control movement. Let:

$n = n_0 + n_1 . 2^1 + n_2 . 2^2 + n_3 . 2^3$ etc.

Phase 2 then is represented by the loop:

while $i < \log_2 N$

    move the contents of processors with $n_i = 1$ left 2^i places.

    $i = i + 1$

    end while.

This method of implementing Phase 2 is especially efficient on computer organisations with 2-dimensional or higher dimensionality connectivity and is indeed general enough to work for the type of movements needed to support the top-down as well as the bottom up algorithms.

References

[1] D.E. Knuth, The Art of Computer Programming, Vol.3 Sorting and Searching Addison-Wesley, Reading, Mass (1973).

[2] C.D. Thompson, and H.K. Kung, Sorting on a mesh-connected parallel computer, CACM 20 (4) 263-271 (1977).

[3] K.E. Batcher, Sorting Networks and their Applications, Proc. AFIBS 307-314 (1968).

[4] G. Baudet and D. Stevenson, Optimal Sorting Algorithms for Parallel Computers, IEEE, Trans. Comput. C-27(1) (1978).

[5] P.M. Flanders and S.F. Reddaway, Sorting on the DAP, Parallel Computing '83, 247-452, North Holland (1984).

[6] F.S. Wong and M.R. Ito, Parallel Sorting on a Re-circulating Systolic Sorter, Computer Journal 27, 260-269 (1984).

[7] M. Stone, Parallel Processing with the Perfect Shuffle, IEEE Trans. Computers, C20, 153-161 (1971).

[8] D. Parkinson, Practical Parallel Processors and Their Uses, Parallel Processing Systems, ed. D.J. Evans, 215-235, Cambridge Univ. Press, (1982).

[9] A.D. Falkoff, Algorithms for Parallel Search Memories, JACM 9, 488-511 (1962).

POSSIBLE APPLICATIONS OF ARRAY PROCESSORS FOR THE PROBLEM OF IMAGE DATA COMPRESSION

Lutz Arnold[1)]

1. INTRODUCTION

In the last few years, the significance of image information within scientific and technical processes has strongly increased, last not least as a result of technological progress in the field of computers and components. To deal with images means to deal with a large amount of data. Despite of the tendency to relatively reduced prices for storage and transmission capacity, there is a constraint for an effective coding of the data. But in this case the point is primarily an effective coding in the sense of transmission and storage of data, not in the sense of a refinement for the later interpretation by men or machine. To minimize the required storage or transmission capacity, a reduction of irrelevancy and redundancy is aspired. In other words, only relevant and nonredundant data should be transmitted or stored to obtain with a minimum of data a maximum of quality at the receiver.
To realise a high compression factor, the corresponding coding algorithm is generally expansive and it increases heavily with the aspired compression. The execution has to be completed in a short time - often in a very short time. So there is a demand for a high calculation capacity. The coding system should be at the same time volume and cost saving.

For data compression, there exists a wide spectrum of applications (see Fig. 1). The requirements concerning the compression in the separate cases are differing partially, but there are common theoretical basics and also in practice there are trends towards an integration.
The author is working in the field of still picture communication with relativ low resolution. For example, a system has been developed, which enables the transmission of an image via telephone line within one minute. The compression is realized by a microprocessor and is exclusively redundancy reducing, that means, identical replication of the image at the receiving side is possible.

---

1) TH KARL-MARX-STADT, Sektion Informationstechnik
9010 Karl-Marx-Stadt, GDR
PSF 964

- Digital TV
  * original data rate 70 Mbit/s (black&white)
    100 Mbit/s (color)
    higher (hi-fi, studio TV)
  * transmission via 34 Mbit/s channel
  * frame frequency 50 c/s
  * HIGH DATA RATES

- Video Telephone (Low Bit Rate Television)
  * transmission via 2 Mbit/s ... 64 Kbit/s ... ... 9,6 Kbit/s
  * transmission time per image 1/25 s ... 2 s
  * LOW CHANNEL CAPACITY

- Still Picture Communication
  * transmission of images of different resolution
  * channel capacity 64 Kbit/s ... 1,2 Kbit/s
  * transmission time 1 s ... 3 min

- Facsimile
  * ISO A 4 documents, 7.7 lines/mm
  * transmission via telephone line or public data network
  * transmission time approximately 1 min
  * HIGH RESOLUTION, BINARY / GREY SCALE

- Digital Storage of Documents
  * weather maps for example
  * LARGE AMOUNTS OF DATA

Fig. 1: Fields of Application for Data Compression

2. PRINCIPLES OF DATA COMPRESSION

Data compression can be obtained by two basic methods:

REDUCTION OF IRRELEVANCY	REDUCTION OF REDUNDANCY
- irreversible modification of the image, loss of information	- reversible modification of the data structure, no loss of information

- decision about relevancy/irrelevancy depends on the special application
- examples
    * definition of the resolution (quantization)
    * elimination of noise and interference
    * elimination of unessential contents

- independent of the application
- exploitation of a priori information about the image, which is of statistical nature

Fig. 2 shows the different phases of an image transmission system (here not only a spatial transmission is meant, a transmission via time is included).

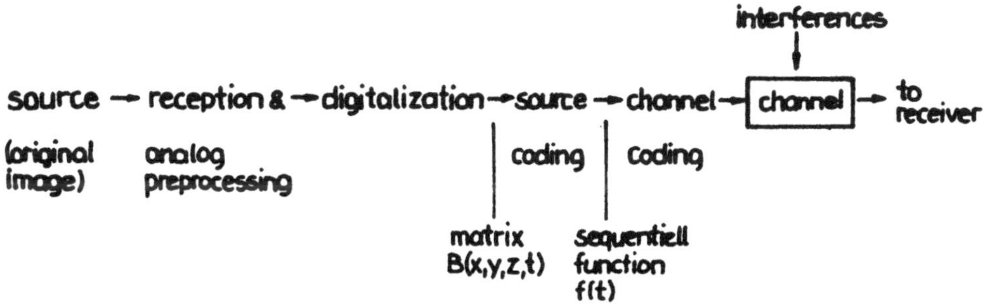

Fig. 2: Different Phases of a Transmission System (Only Tranmitter)

The process of data compression is restricted to the phases between reception and source coding. In later phases (especially at the receiver), compression is not obtained. Of special interest in this paper is the source coding process and for that reason it should be further detailed (see Fig. 3).

Segmentation & Serialization

The intention is to catch the spatial and temporal relations inside the image. That means, a model of the statistical properties and dependencies must be formed. From this the definition of a new alphabet of image elements is derived. This can also be called a reinterpretation of the image (f. e. the complete image is not seen as a sum of pixels, but as a sum of blocks or runs). The probabilities of occurance of the image elements should be spread (close to 0 or 1) and from the knowledge of already scanned elements should be possible. So the entropy of the choosen statistical model is very low and the acquired redundancy is high. The image matrix is at this occasion serialized.

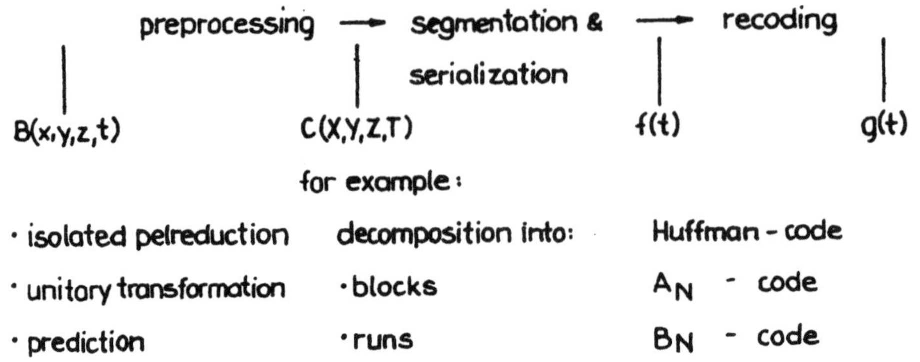

Fig. 3: Source Coding

Recoding

It is optimally realized by the Huffman Code, which transforms elements of higher probability into short code words and elements with a low probability into long code words. Also subobtimal, easier managable and systematically producable codes are used. This phase eliminates the previously acquired redundancy.

Preprocessing

It is aiming at an improvement of the statistical properties (removal of interferences or unimportant details) or at a transformation of complex relations inside the image into simpler ones. It is leading to a higher efficiency and to simpler algorithms in the following phases. Product of the preprocessing is a matrix of the same size as the original matrix, but the semantics of the coordinates can be changed.

Especially the operation, which takes a matrix and produces a matrix, is of interest for array structures. It can be parallelized in high degree at microlevel and the possible gain of time is immense. The processing of the later phases can also be parallelized (pipelining at macrolevel), but only in lower degree. By an optimization of the preprocessing, a disburdening of the other, less parallelized phases can be attained and there a reduction in execution time is the result.
Therefore the preprocessing can algorithmically be the most extensive process.

In source coding parallelization is possible on micro- as well as on macrolevel without an embarrassing extent of organisation overhaed.

3. IMPORTANT PREPROCESSING METHODS AND CORRESPONDING ARRAY STRUCTURES

3.1. UNITARY TRANSFORMATIONS

The whole image is devided into blocks, which are transformed in that manner, that

- the individual elements of the transformed image matrix are decorrelated and can be coded isolated,
- the variance of the elements of the transformed matrix is reduced (compression by following requantization and Huffman Coding),
- a maximum of energy is concentrated in a minimum of elements (compression by suppression of nonessential elements).

That means, a compression is achieved by irrelevancy and redundancy reduction.
The operation to be performed can be formulated generally:

$$Y = A\,X \qquad (1)$$

X - original image matrix (one block),
Y - transformed matrix (one block),
A - transformation matrix.

Written in another way:

$$y_{i,j} = \sum_k \sum_l a_{i,j;k,l} x_{k,l} \qquad \text{(two dimensional blocks)} \qquad (2)$$

$$y_i = \sum_k a_{i,k}\, x_k \qquad \text{(one dimensional blocks)} \qquad (3)$$

Wanted are these transformations, which are fullfilling the demands set up above. The optimum transformation is the Karhunen-Loeve-Transformation, but the transformation matrix A is a function of the block to be transformed X and it's determination is very expansive. In practice the Cosine-, Slant-, Haar- or Hadamard-Transforms are used (a special advantage of the Hadamard-Transform is that it's coefficients $a_{i,j;k,l}$ are only +1 or -1; so no multiplication has to be performed).
A universal structure, suitable for all transformations would be desirable, but a restriction to a special transformation (f.e. to the Hadamard Transform) can bring an essential structure simplification.
Fig. 4 shows an overview of an transformation processor. The heart is an ALU-Array for a parallel generation of the elements $y_{i,j}$ (see equ. (2)). The DCP serves for the distribution of the operants $x_{k,l}$ and $a_{i,j;k,l}$ to the

ALU cells, for the collection of the results y_{ij} and for the control of the ALU array.

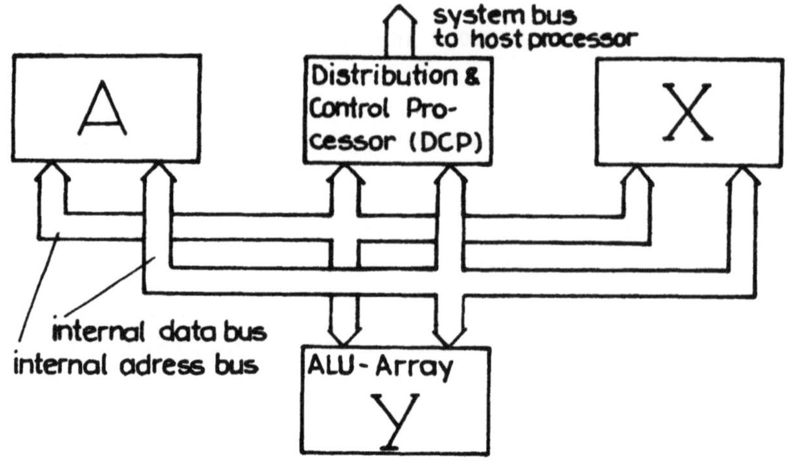

Fig. 4: Transformation Processor

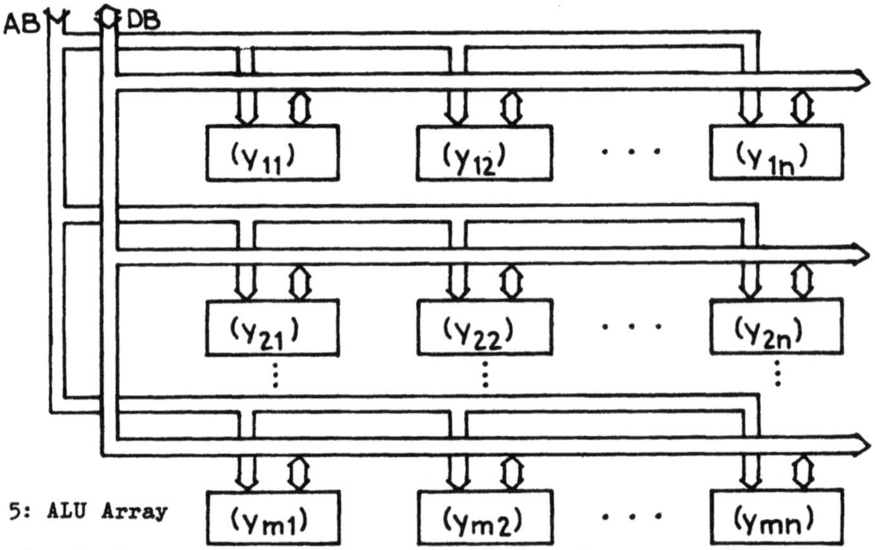

Fig. 5: ALU Array

The individual components can exist separately (the DCP can be represented by the system CPU, the matrices X,A can be stored in the normal workspace memory and the ALU array can exist as coprocessor). More desirable is an integration of the complete structure on one chip, which works as a subprocessor.

Fig. 5 shows the ALU array and Fig. 6 the structure of one individual ALU cell. The individual cell can be more or less complex (f.e. hardware realization of multiplication or only summation and shift), thus the ge-

generation of a transformed element y is devided into more or less detailed steps of higher or lower logical level.

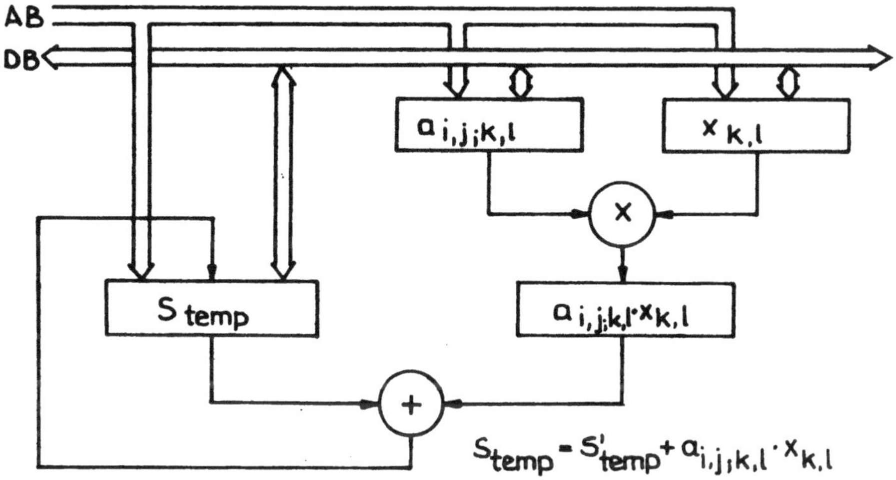

Fig. 6: Individual ALU Cell

The usual block sizes (or the necessary array sizes) are ranging from 4 x 4 to 16 x 16 elements.
Because the blocks are transformed individually, a parallelisation can also be performed at block (macro) level.

3.2. Prediction

This method uses the dependencies between the pixels to predict the value of the pixels from the knowledge of the value of other pixels.
The prediction value $x'_{i,j}$ of a pixel $x_{i,j}$ is given by:

$$x'_{i,j} = \sum_{m}\sum_{l} b_{i,j;m,l}\, x_{m,l} \qquad \begin{matrix} i \neq m \\ j \neq l \end{matrix} \tag{4}$$

The transformed pixel is given by the deviation between prediction and real value of the pixel:

$$y_{i,j} = x_{i,j} - x'_{i,j} \tag{5}$$

From equ. (4) and (5) follows:

$$y_{i,j} = \sum_{m}\sum_{l} c_{i,j;m,l}\, x_{m,l} \qquad \begin{matrix} c_{i,j;m,l} = -b_{i,j;m,l} \\ c_{i,j;i,j} = 1 \end{matrix} \tag{6}$$

The equations (2) and (6) are formally identical and therefore the same array structure is usable for this type of transformation. But the prediction has special properties and conditions, which cause vigorous structural simplifications.

Fig. 7 illustrates the situation during the prediction of one pixel. For a later retransformation and regeneration of the original image at

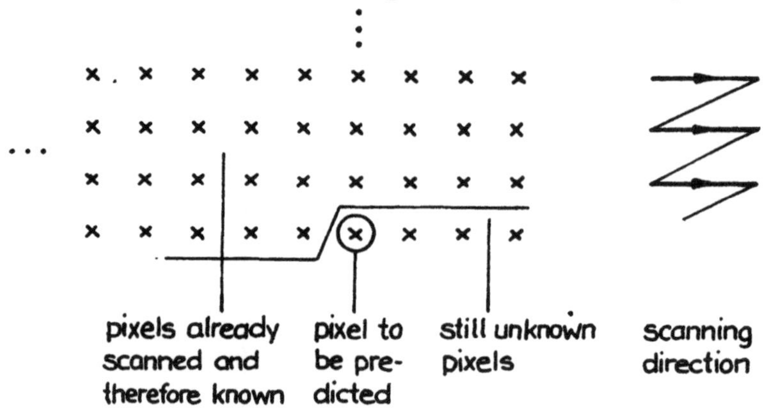

Fig. 7: Prediction Relevant Pixels

the receiver , the causal principle has to be used. That means, the derivation of the prediction is only allowed of already scanned pixels (which are at the retransformation of the image already regenerated). The consequence is, the transformation can be parallelised but not the recursive retransformation. Moreover, the usable coherences and relations between the pixels are in practice spatially very limited and often spatially invariant. All that means that many of the elements of the transformation matrix in equ. (6) are zero and the number of differing coefficients is limitted. A dropping of the matrix multiplication model is usefull.
Fig. 8 shows some usual and possible predictors (that are configurations of relevant pixels for the prediction).
Often only the neighbouring pixels are relevant and only specific and synthetic image contents (f.e. layouts) can have more far-reaching coherences. This has a positiv consequence for the implementation of an array. The preferred configuration is shown in Fig. 8 c). The inclusion of more pixels has hardly an effect.

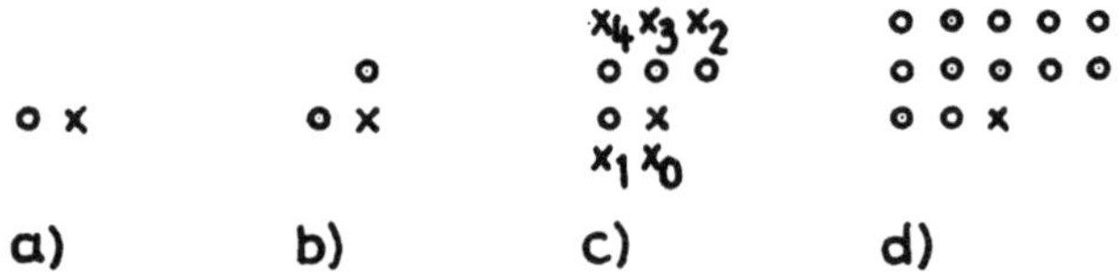

x - pixel to be predicted; o - relevant pixel

Fig. 8: Usual and Possible Predictors

Fig. 9 illustrates a desirable array arrangement. It shows also the operation to be performed by each cell. The coefficients can be selected as constant, but they are depending on the image type to be processed. One possible and often valid example for this operation is

$$y_0 = x_0 - \frac{x_1 + x_2 + x_3 + x_4}{4} \qquad (7)$$

The values y can additionally be transformed on the grey scale (non negative values).

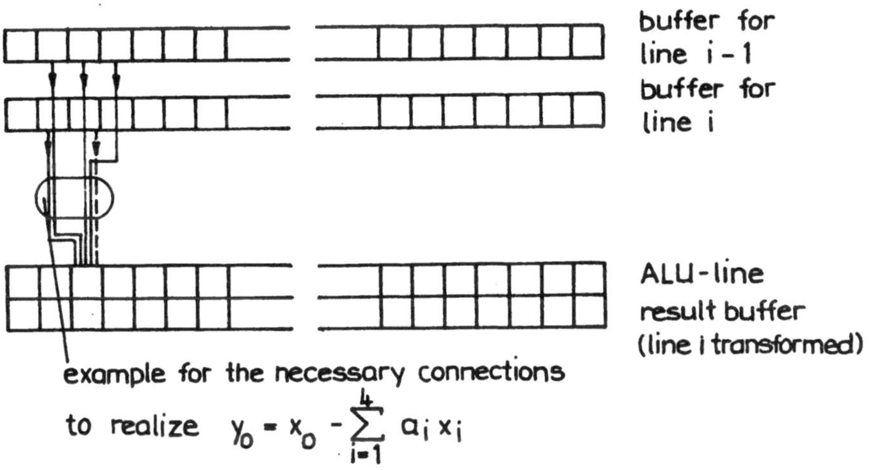

Fig. 9: Proposal for a Prediction Array

The pixels of the produced differential image are within the known and used coherences decorrelated and the image can be coded effectively by simple coding schemes (f.e. one dimensional run length coding).
The prediction is an exclusively redundancy reducing method.

3.3. Image Enhancement

Image enhancement in connection with data compression serves to remove image ingredients, which are worsening statistics. This irrelevancy reducing technique leads to a decrease in image entropy and increases for example the average run length or the number of pure blocks. Primarily the intention is to abolish disturbances or unnecessary details.
There are several methods to process adjoining pixels. Some basic pixel configurations ("operators") are shown in Fig. 10.
Here are only some possible operations:

- Isolated Pel Reduction.
  The pixel is eliminated, if there exists no pixel of the same value in it's neighbourhood.

- Treshold Method.
  The pixel will be eliminated, if the pixel has less than k pixels of the same value in it's neighbourhood.

- Averaging Method.
  The pixel is replaced by the average value of all pixels in the operator. The replacement can depend on special conditions.

The operator configuration can vary in size and the definition of the neighbourhood relations can be more generalized.
Even if the special operation to be performed is differing from prediction, also the immediately neighbouring pixels are processed. So it is possible to design a structure, which is able to realize line by line the prediction as well as the image enhancement. These two methods can be combined in one coding process and they can be performed in time sharing by one array or in pipelining by different array processors.

x - pixel to be modified o - relevant pixel

Fig. 10: Basic Configurations of Image Enhancement Operators

4. Conclusions

Image coding offers a series of possibilities for parallelization of the algorithms. Three principle methods and their intentions (Unitary Transformation, Prediction, Image Enhancement) are explained.
Integrated array structures enable the parallelization of the coding process in high degree for practical use. They are supporting the increase in computation power, which is necessary especially for the processing of image sequences (processing of complete images or many thousands of KBytes within fractions of seconds). The application of arrays permits the exploitation of redundancy and irrelevancy in a degree not possible with usual processors.

References

/1/ Netravali,A.N.; Limp,O.J.:
Picture Coding: A Review.
Proc. IEEE 68, March 1980, pp. 366.

/2/ Jain,A.K.:
Image Data Compression: A Review.
Proc. IEEE 69, March 1981, pp. 349.

A CELLULAR RECONFIGURABLE DATA FLOW COMPUTING SYSTEM

Masateru HARAO[1] and Paisan LOMTONG[2]

Abstract This paper describes a multiprocessor system capable of directly and efficiently executing data flow graph languages. The proposed cellular reconfigurable data flow computing system (abbreviated to CECODA) consists of two uniformly interconnected networks of processing units, one of which is a linear (or rectangular) array of identical elements called the active cells, and the other a 2-dimensional (3-dimensional) cellular array of identical elements called the communication cells. Both kinds of cells have modest processing and storage requirements. We describe first an algorithm to transform a given program written by certain high level language to a data flow graph program and a formal definition of the data flow graph language language and its semantic are studied. The data flow graph program is mapped homeomorphically into cellular processor such that each node corresponds to an active cell and each edge to a path. An efficient graph mapping algorithm into CECODA is studied precisely. A parallel routing and switch setting algorithm of cellular interconnection array are given also.

1. Introduction

Recently VLSI technology makes available to use large number of inexpensive processors , creating the capacity to do computation which previously was not possible. Accordingly it seems possible to construct a cellular type computing system which is composed lots of, possibly 10^5 or more, cells. One of such examples is the systolic array and hardware algorithms[2]. On the other hand, there also exist some attempts to construct a system which is able to realize a certain wide class of hardware algorithms dynamically by introducing some programmable mechanisms[3,4]. The cellular reconfigurable data flow computer architecture(CECODA) proposed in this paper is based on fundamentally the similar motivation. The programs of CECODA are described as a directed graph with token label like usual data flow mechanisms, where each node corresponds to a program module which is computed by a processing cell, and each arc shows a data path. The program is mapped directly into the cellular array homeomorphically in such a way that each node is mapped to a processing cell and each arc to a path composed the communication cells. By this architecture we are able to expect a highly parallel and epochal machine. We shall overview the outline of this CECODA architecture.

2. Functional Graph and Data Flow Computation

2.1. Functional graph

A functional graph G is a directed graph with labelled nodes and arcs. The label set for nodes is denoted L. In formally, L is a finite set of the following symbol set:

i) n-ary function symbols: $f_i^{(n)}$ $(i \geq 1,\ n \geq 1)$,

ii) n-ary predicate symbols: $p_i^{(n)}$ $(i \geq 1,\ n \geq 1)$,

iii) Fork operation symbols: $\sigma_i^{(n)}$ $(i \geq 1,\ n \geq 2)$,

iv) Gate operation symbols: Ⓣ, Ⓕ ,

v) Branch operation symbols: Ⓑ ,

vi) Merge operation symbols: Ⓜ ,

vii) Joint selector operation symbol: JS .

The labels of arcs mean the communication

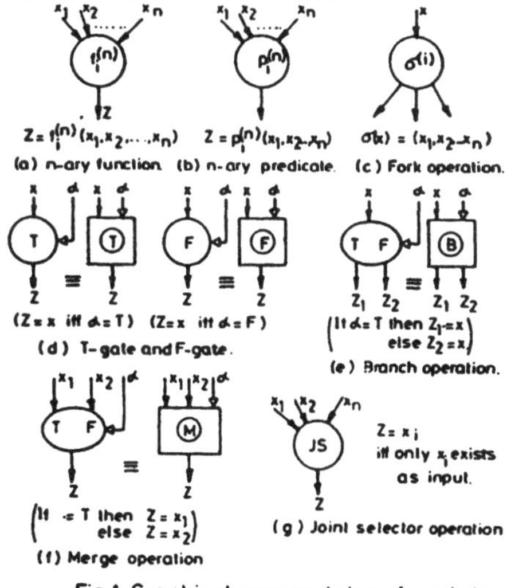

Fig.1. Graphical representation of symbols.

1) Faculty of Engineering, Yamagata University, Zyounan 4-3-16, Yonezawa, 992 JAPAN.
The work of M.Harao was partially supported by Science Research Grant of Japan, (c) 59580015.

2) Research Institute of Electrical Communication, Tohoku University, Katahira 2-1-1 Sendai, 980 JAPAN.

channel through which the inputs and outputs are transmitted. We use the two types of variables, i.e., the data variables and the control variables.

i) Data input variables: $X = \{x_1, x_2, \ldots.\}$,

ii) Data output variables: $Z = \{z_1, z_2, \ldots.\}$,

iii) Control variables: $\Gamma = \{\alpha_1, \alpha_2, \ldots..\}$.

The graphical representation of the nodes is illustrated in Fig.1 and their intuitive meaning are also described under their graphs.

2.2. Processor network and Firing rule

The nodes and the arcs of the functional graphs represents the program modules to be computed and the flows of data respectively. We assume that a program module corresponding to a node of the functional graphs can be executed by processing unit. If we restrict the processing units to be the simple ones such as adder, multiplier, subtractor and so on, the the functional graphs may be considered to be the hardware algorithms. On the other hand, if we take the ordinary computer as processing unit, then the functional graphs may be considered to be the computer networks.

We assume here that the processing unit is a small size general purpose microprocessor such that hundreds of it can be manufactured on one chip. The processing units are interconnected by a certain interconnection network and we call it a processor network. As for the computation by processor network, we assume the following conditions.

1) At the function program module level, the computation is executed under the data flow control.
2) At the low level (inside the program modules), the computation is executed under the control flow.

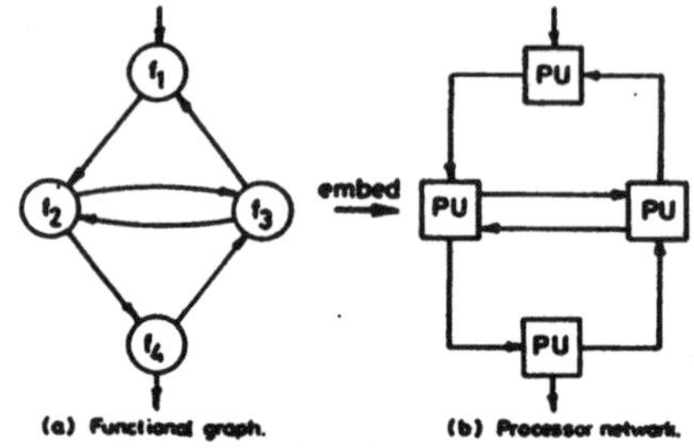

(a) Functional graph. (b) Processor network.

Fig.2. The relation between Functional graph and Processor network.

We introduce the two kinds of firing rules at the functional program module level.

1) General firing rule:
   If all the inputs are defined,
   then fire else wait.
2) Extended firing rule:
   If a subset of the inputs is defined and there exists a submodules of the program module which is firable by the input set,
   then fire else wait.

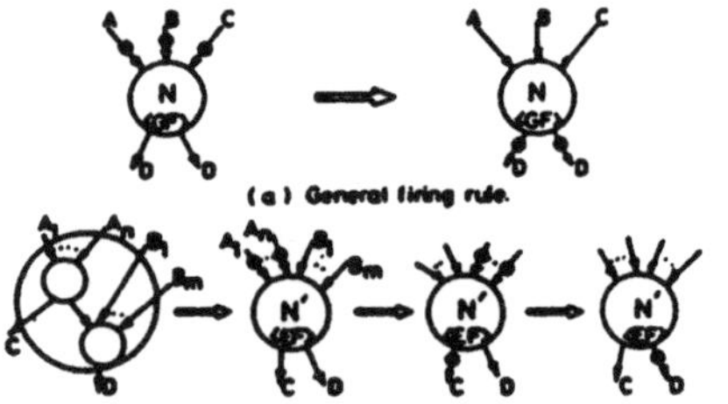

(a) General firing rule.

(b) Extended firing rule.

Fig.3. Firing rules.

The rough sketch of this is illustrated in Fig.3. By this observation, we can look the functional graph as the data flow graph and we often use the data flow graph instead of the functional graph if no confusion arises.

3. CECODA Language

3.1. Design principle

There are several approaches to design the user language suitable for such data flow computation by the processor networks. One approach is to design the new graphical data flow language and the other is to use the conventional language like FORTRAN by constructing a compiler which transforms it into a graph language. Considering the utilizability of the software resources developed(until now), we adopted the second approaches. The outline of the design principle is sketched in Fig.4.

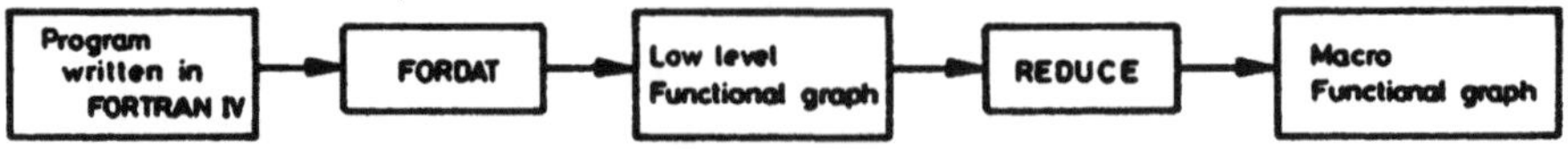

Fig.4. Design principle of CECODA language.

3.2. FORDAT Procedure

The key ideas of FORDAT are summarized as follows.

1) partition a FORTRAN IV program into subblocks such that each subblock contains only one transfer-in control variable at the first statement and only one transfer-out control variable at the last statement.

2) Realize the control flow mechanism as a data flow graph among the subblocks using the JS node.

The details of the FORDAT procedure should be referred to [5].

Theorem 1. The time complexity of FORDAT procedure is $O(m^2n)$ and the number of the transformed graphs are $O(mn)$, where m is the number of statements and n is the number of the variables of the longest statement.

3.3. REDUCE procedure

Three graph reduction rules are proposed here.

1) The first graph reduction rule is for merging several control nodes and gate nodes into a processing node(see Fig.5(a)).

2) The second graph reduction rule(Fig.5(b)) is for merging several processing nodes operating under the general firing rule into a macro processing node operating under the general firing rule.

3) The third rule(Fig.5(c)) is the extension of the second rule, and it is required when the number of nodes is still greater than the number of processing units even if the second rule is applied.

The detail of the graph reduction algorithm should be referred to [6].

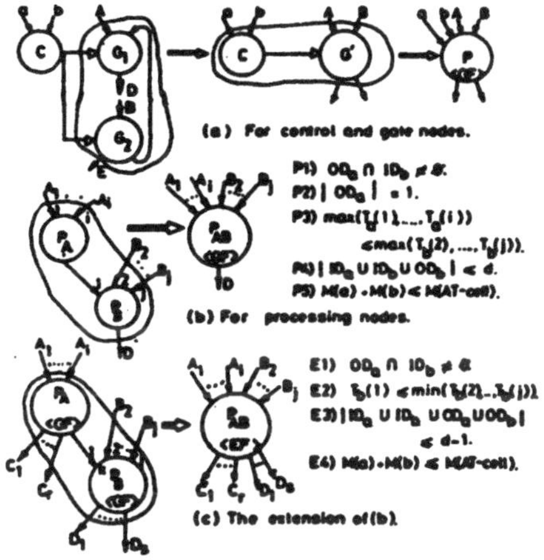

Fig.5. Reduction rules.

Theorem 2. The time complexity of REDUCE algorithm is $O(mnd^2)$ and the number of nodes of the merged data flow graph is $k(mn)$ where $\min(D/d, 1/M) \leq k \leq 1$, D and M are the number of I/O ports and the memory area of processing units, and mn and d are the order and degree of the given data flow graph.

Example 1. Polynomial evaluation.

For evaluating the polynomial P(x) of degree n, $P(x) = a_0 + a_1x + a_2x^2 + \dots + a_nx^n$, we shall propose an algorithm which is a modification of the algorithm introduced by Dorn(7). By this we can evaluate P(x) in parallel and is described as follows:

$$x_i^{(0)} = x;\quad c_i^{(0)} = a_i;\quad i = 0,1,\dots,n.$$

$$x_i^{(1)} = x_i^{(0)} * x_{i+1}^{(0)};$$

$$c_i^{(1)} = c_i^{(0)} + c_{i+1}^{(0)} * x_{i+1}^{(0)};$$

$$i = 0,2,\dots,2\ n/2.$$

$$\cdots\cdots\cdots$$

$$x_i^{(m)} = x_i^{(m-1)} * x_{i+2m}^{(m-1)};$$

$$c_i^{(m)} = c_i^{(m-1)} * c_{i+2m}^{(m-1)} * x_{i+2m}^{(m-1)};$$

$$i = 0, 2m+1, \dots, (2m+1)\ n/(2m+1).$$

$$P(x) = c_0(m).$$

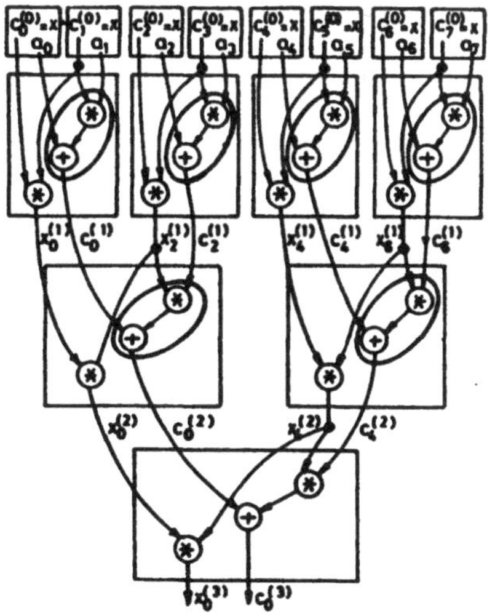
Fig.6. The data flow graph of p(x) of degree 7.

The following is a program of this.

```
C EXAMPLE POLYNOMIAL
      DIMENSION X(20,20), C(20,20)
      READ(5,1) N, M
    1 FORMAT (I4)
      READ(5,2) (A(I), I = 1,N), X1
    2 FORMAT (F10.3)
      DO 3 I = 1,N
```

```
      X(I,1) = X1
    3 C(I,1) = A(I)
      DO 4 J = 1, M
      S = 2/(J+1)
      DO 4 I = 1,N,S
      X(I,J) + X(I,J-1) * X(I+2J,J-1)
    4 C(I,J) = C(I,J-1) + C(I+2J,J-1) * X(I+2J,J-1)
      WRITE(6,2) C(1,M)
      STOP
      END
```

The data flow representation of this algorithm is given by the FORDAT algorithm as illustrated in Fig.6. By applying the REDUCE procedure we can merge the * and + nodes, where the merged nodes are encircled by the bold lines. Further, if we merge the two nodes enclosed by a box as a module, then a tree structured macro data flow graph is obtained, where a module has to be executed by a processing unit.

4. Interconnection Networks for Processor Network

4.1. System Architectures

In order to compute the data flow graph program, we use the processor networks which can be reconfigured according to the program structures. Roughly speaking we can classify the structures of processor networks into two classes, the static one in which processors are rigidly interconnected in some regular patterns like cellular arrays and the dynamic one in which processors are combined by certain interconnection network[8,9,10]. To reconfigure the processor networks according to any input data flow graphs, the static models are not suitable since the reconfiguring algorithm is too difficult and we can not expect the efficient utilization of hardware resources. Therefore we adopt the dynamic model.

4.2. Interconnection Network

As the discipline of the system architecture, we consider the following factors;

1) Universality of the system concerning with the graph embedding,
2) Simple cell organization,
3) Existence of efficient and simple algorithm for reconfiguring the processor network.
4) Realizability of efficient data routing mechanism.

When we discuss the network architecture, it is important to consider the following factors.

1) Operation mode;
   Synchronous or Asynchronous.
2) Switch cell control mode;
   Centralized or Distributed.
3) Switching methodology;
   Packet switching or Circuit switching.
4) Network structure;
   Universal or Special purpose.

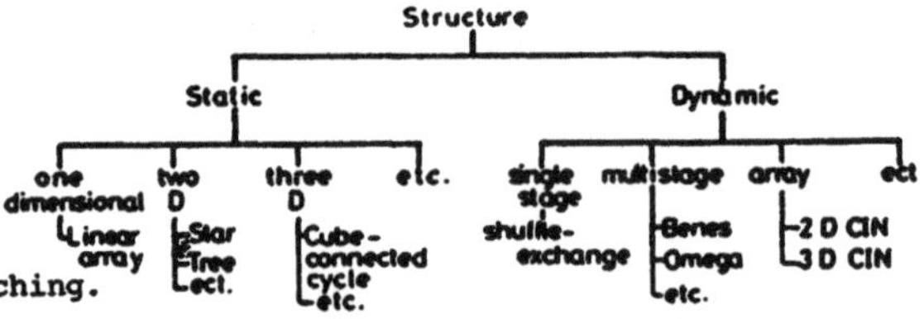

Fig.7. Structures of interconnection networks.

A distributed switch setting mechanism for routing is called self-routing if each switch cell determines its own setting depending only on the preset incoming routing data[11]. It seems essential for the interconnection network of CECODA systems to have a facility of a synchronous operation, distributed switch cell control , universal network structure and the self-routing ability. From such viewpoints, we shall propose the cellular interconnection networks(CINs) which are universal concerning with realizing any interconnection patterns among processors.

5. Processor Network Realization

5.1. Basic concepts

We shall discuss the processor network realization algorithm by CIN assuming the following conditions.

1) Each SW-cell can only be interconnected with its direct neighbor cells.
2) No overlapping among connection lines is allowed(duplicated use of the same port is not allowed).
3) Crossing among connection lines are not allowed.

Definition 1. An embedding $f = (f_1, f_2)$ of $G = (V,E)$ in $\mathbb{C}_k = (C_k, S_k)$, $i = 1,2$, is a mapping defined as follows:

1) f_1 is an injection from $V \times \{1,..,d\}$ into $C_k \times \{1,..,d\}$ called the cell-port assignment, where d and p are the degree of G and the number of ports of AT-cell respectively.
2) f_2 maps E into the path of C_k such that every pair of such paths is edge disjoint.

The procedure to determine an embedding mapping (f_1, f_2) is called the set-up algorithm and the procedure to establish the interconnection in CIN using the data obtained by the set-up procedure is called the routing algorithm.

5.2. Routing algorithm

5.2.1. Routing patterns

The routing patterns should be devised so that the packet switching or circuit switching setting of CIN can be performed effectively using simple routing data. Taking account of such factors we adopt the routing patterns given in Fig.9.

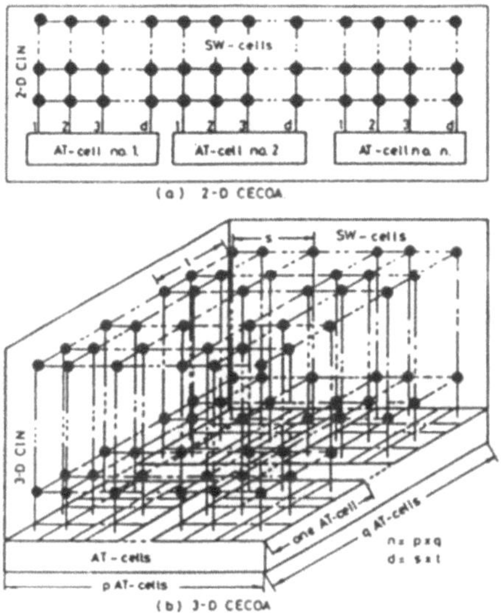

Fig.8. CECOA Systems.

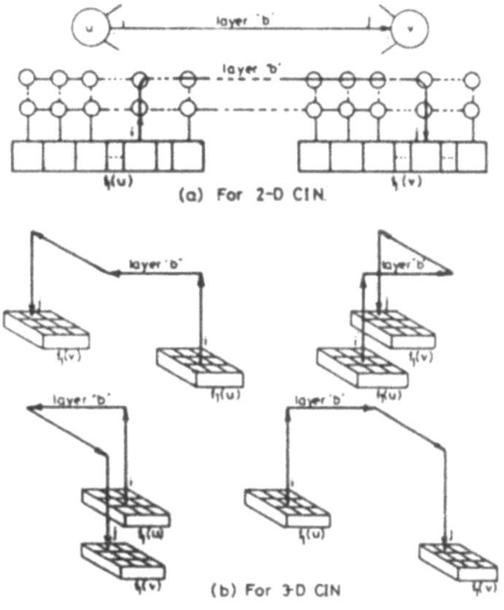

Fig.9. Routing patterns in CIN.

(a) For 3-D CIN

3 bits d_i Direction	log m bits c_i Steps of routing
0 0 1	b-1
0 1 0	b-1
0 1 1	$X_0 - X_i - 1$
1 0 0	$X_i - X_0 - 1$
1 0 1	$Y_0 - Y_i - 1$
1 1 0	$Y_i - Y_0 - 1$

(b) For 2-D CIN

2 bits d_i Direction	log m bits c_i Steps of routing
0 0	b-1
0 1	b-1
1 0	$X_0 - X_i - 1$
1 1	$X_i - X_0 - 1$

Fig.10. Routing data.

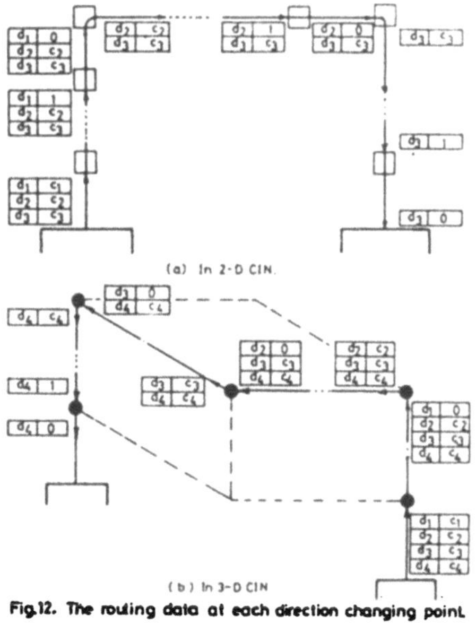

Fig.12. The routing data at each direction changing point.

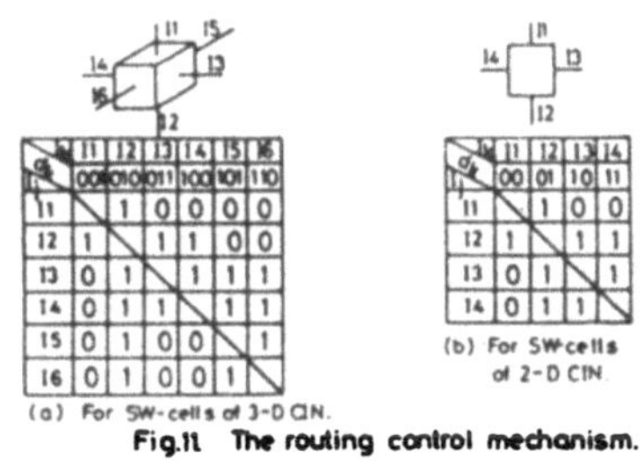

Fig.11 The routing control mechanism.

5.2.2. Logical behavior of SW-cells

The logical operation of each SW-cell is designed as in Fig.11, where the I_i row, I_k column component means that I_i is connected with I_k by input k iff its component is 1.

5.3. Pattern Setting Mechanism

Since we use only simple routing patterns of Fig.9, there are at most three direction changing points of the routing patterns in 3-D CIN and at most two direction changing points in 2-D CIN. To establish a simple routing, we devise the step count down method.

1) Total routing data consists of 4 (or 3) data packets for 3-D (2-D) CIN. Each routing data is composed of the direction bits and the binary representation of the path length of the routing for this direction.

2) When a routing data arrives at a SW-cell, the routing path length bits are decreased by one if it is not zero or the direction change will be done if it is zero.

This is illustrated in Fig.12.

Theorem 3. The size of the required routing data is $4(\log(m^{1/2}) + 3)$ bits for 3-D and $3(\log m + 2)$ bits for 2-D respectively, where m is the total number of ports of CIN.

5.4. Cell-port Assignment

In order to reduce the required layer of CIN, we distinguish the ports of each cell to out-port and in-ports. But this restriction is not essential because the communication of both directions can be done by constructing a physical communication line.

Roughly speaking the cell-port assignment consists of following procedures:

1) Transformation to digraph: This procedure transforms a given graph to a digraph by assigning directions to each edge.
2) Cell assignment: This procedure defines the mapping f_1 from node set V to the AT-cell set.
3) Port assignment: This procedure defines the mapping f_2 from $V \times \{1,\ldots,d\}$ to $f_1(v) \times \{1,\ldots,p\}$.

Proposition 1. Let G is an arbitrary graph of degree d and of order n. Then G can be transformed to a digraph G^* such that if d is even, then the outdegree and the indegree of each node are both d/2, and if d is odd, then one of them of each node is (d+1)/2 and the other is (d-1)/2. The algorithm take O(dn) time and O(dn) space.

The formal definition of cell-port assignment $f = (f_1, f_2)$ is given as follows:

a) f_1 is an injection from V into C'_k, k = 2,3, such that
   $C'_2 = \{(x,0);\ 0 \leq x \leq n-1\}$,
   $C'_3 = \{(x,y,0);\ 0 \leq x,y \leq m\}$, where n is the order of V and $m = \lceil n^{1/2} \rceil$.

b) f_2 is an injection from $V \times \{1,\ldots,d\}$ into $C'_k \times \{1,\ldots,p\}$, k = 2,3, such that
   (i) $f_2((v,i))$ is an out-port of $f_1(v)$ if (v,i) is an out-arc of v,
   (ii) $f_2((v,i))$ is an in-port of $f_1(v)$ if (v,i) is an in-arc of v.

The cell-port assignment mapping f is the composition of f_1 and f_2, i.e., (f_1, f_2) defined as follows: $(f_1,f_2)((v,i)) = (f_1(v), f_2(v,i))$ for any $(v,i) \in V \times \{1,\ldots,d\}$.

Example 3. Cell-port assignment for binary trees. Let G be a n-leaf binary tree graph. We define the cell-port assignment in the following ways.

(1) 2-D case

a) Divide Divide G into k-level, where $0 \leq k \leq h = \lfloor \log n - 1 \rfloor$. WE assume that the leaf nodes and the root node are located at k = 0 and at level h respectively
   i) $f_1(v_{root}) = 2^h$.
   ii) For a node v at level k and i-th branch,
   $$f_1(v) = 2^k(2i-1), \text{ where } 0 \leq k \leq h, \text{ and } 1 \leq i \leq 2^{h-k}.$$
   To illustrate this correspondence, we show a binary tree labelled by the above mentioned algorithm in Fig.13(a), where the node labelled i means that it is mapped to the processor P_i by f_1. In Fig.13(b), an euler partition is given, where the dotted arcs are the added ones.

b) We assume that the AT-cell has the organization illustrated in Fig.13(d), where port 1 is the inport and port 3 is the outport respectively, but port 2 can be used both for the inport and outport.
   We define f_2 as follows: $f_2(v,i) = i$ for i = 1,2,3.

(2) 3-D case

a) Suppose that the nodes of G are labelled by the same integers to 2-D case (Fig.13(a)). Let $n = 2^i$ and i is an arbitrary integer.

i) if i is odd, then we embed G into 2g x q array, where $q = 2^{(i-1)/2}$.

ii) if i is even, then we embed G into q x q array, where $q = 2^{i/2}$.

The mapping f_1 is defined recursively as shown in Fig.14(a). An example is given in Fig.14(b).

b) We assume that the AT-cell has the organization illustrated in Fig.13(e). If we allow to use the port 2 both for the inport and the outport, then we can set the port assignment as $f_2(v_i) = i$.

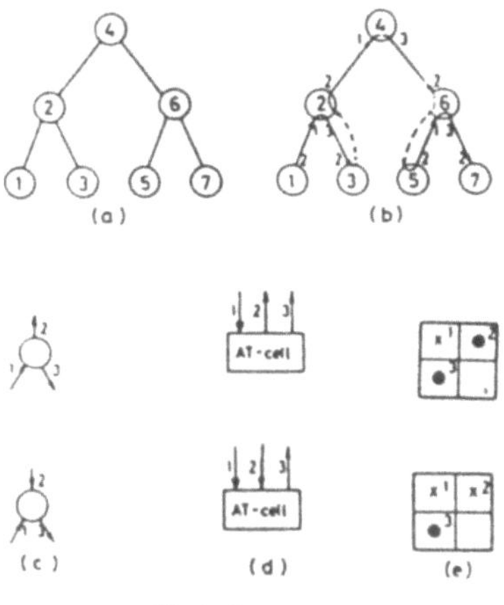

Fig.13. Binary tree

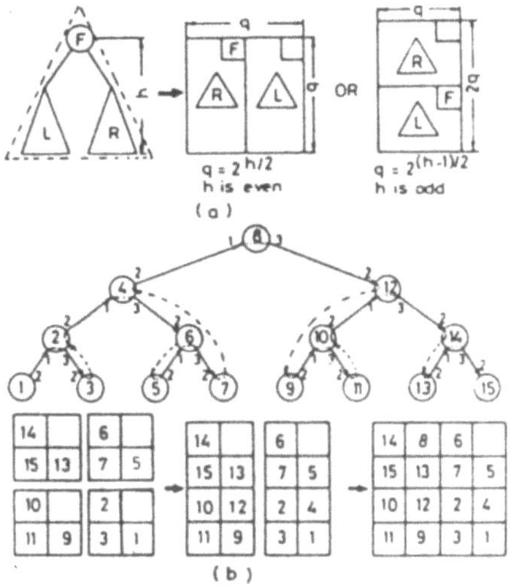

Fig.14. Cell assignment.

6. Polynomial Time Realization Algorithm

We concentrate mainly to the three dimensional case similar method is available for two dimensional case. The detailed discussion should be referred to (14).

6.1. Bipartite Graph Representation

We assume here that the first layer of CIN consists of p x q AT-cells and each AT-cell possesses s x t ports.

A bipartite graph(or simply bigraph) is a graph whose node set V can be partitioned into two disjoint subsets V_1 and V_2 such that every edge of G joint V_1 with V_2. If more than one edge can joint two nodes(multiple edges), then it is called a multigraph. For any given graph G and any cell-port assignment, we construct a bigraph G_B as follows:

Procedure: Bipartite graph construction

Input: A graph G = (V,E); A cell-port assignment $f = (f_1, f_2)$.

begin:

(1) Partition each AT-cell into s x t smaller segments, and associate the points r_j and c_i to the j-th row and the i-th column, where $1 \leq j \leq t \times q$, and $1 \leq i \leq s \times p$. (This is illustrated in Fig.15(a).)

(2) From G we derive a directed graph $G_s = (V,A)$. (Proposition 2)

(3) Define the (multi) graph $G_B = (X,B)$ such that

i) $X = \{\{r_j\} \cup \{c_i\};\ 1 \leq j \leq t \times q,\ 1 \leq i \leq s \times p\}$.

ii) $B = \{\{r_j, c_i\};\ f(u,a) \in r_j,\ f(v,b) \in c_i,\ \{(u,a),(v,b)\} \in A\}$, where a, b are the labels of this edge.

end;

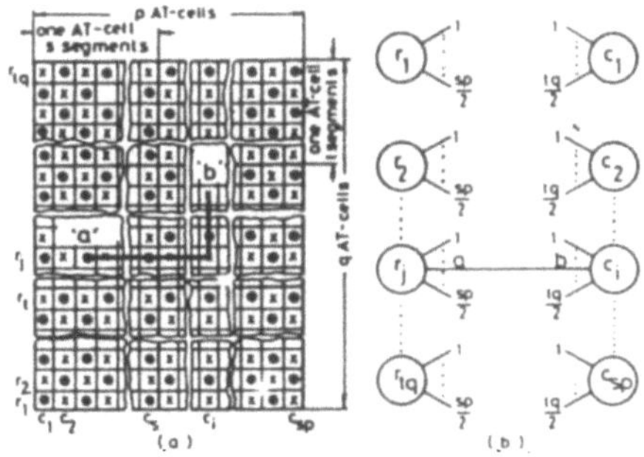

Fig.15. Bipartite graph representation of G.

Proposition 2. The graph $G_B = (X,B)$ is a bipartite graph of order sp + tq and of degree at most max(s/2 .p, t/2 .q). (The size of G_B is the same to G.)

6.2. Layer Assignment Algorithm

An edge-coloring of a graph is an assignment of colors to its edges so that no two adjacent edges are assigned the same color. Generally the problem of edge-coloring belongs to the class of NP-complete problems, but for the class of bipartite graphs the more efficient algorithms are available.

We use the modified algorithm of color-by-partitions[15] and it is noted that the minimum number of required colors is equal to the degree of the bipartite graph.

Lemma 1. Modified color-by-partition algorithm finds a minimum coloring, in time $O(|V|^2 \log|V|)$ and space $O(|E|+|V|)$, where $|V|$ and $|E|$ denote the order and size of G_B respectively.

We assign the edge colored by the b-th layer of CIN. Since the usable routing patterns are restricted to the one of Fig.9, no overlapping arise among interconnections.

Lemma 2. Let G_B be any bigraph. Then G_B can be realized in CIN with out overlapping using 4 routing patterns in Fig.9.

Theorem 4. An arbitrary graph G of degree d and of order n can be realized in CIN using at most $D = \max(\lceil s/2 \rceil .p, \lceil t/2 \rceil .q)$ layers.

6.3. CIN minimization

From the results of previous section, the number of required layers depend on the organization of AT-cell. In order to establish an universal CIN with small layers, we have to construct the bipartite graph as possible as small degree. We shall give next a AT-cell organization which satisfies such requirements

A Latin square is a matrix having integers as its components such that each distinct integer occurs exactly once in each row and each column of it. For designing an AT-cell of d-ports, we use a s x s Latin square, where $s = \lceil d^{1/2} \rceil$.

Roughly speaking we design the ports of even number to the out-arcs and the odd ones to the in-arcs.

If the number of out-arcs(or in-arcs) of certain nodes exceeds d/2, then an extra port not used is to be assigned further(Fig.16).

It is noted that the number of out-ports (or inports) in each column or each row of the matrix is at most s/2, where $s = \lceil d^{1/2} \rceil$.

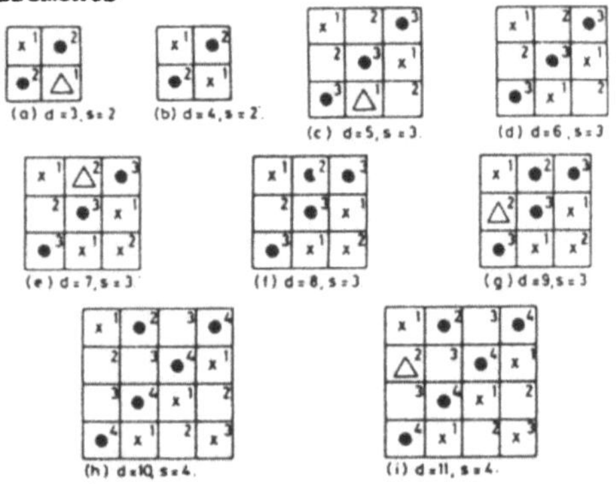

Fig.16. AT-cell organization based on Latin square.

Theorem 5.(Main theorem) An arbitrary graph G of degree d and of order n can be realized in CIN with $D = (1/2)(dn)^{1/2}$ layers and therefore with $(1/2)(dn)^{3/2}$ SW-cells. The realization algorithm runs in $O(dn.\log(dn))$time and $O(dn)$ area.

Example 3. We embed the graph given in Fig.17(a). First we define the cell-port assignment $f = (f_1, f_2)$ and is given in Fig.17(b). Under this mapping f, the bigraph G_B is obtained as shown in Fig.17(c). Thus we have the interconnection of Fig.17(d).

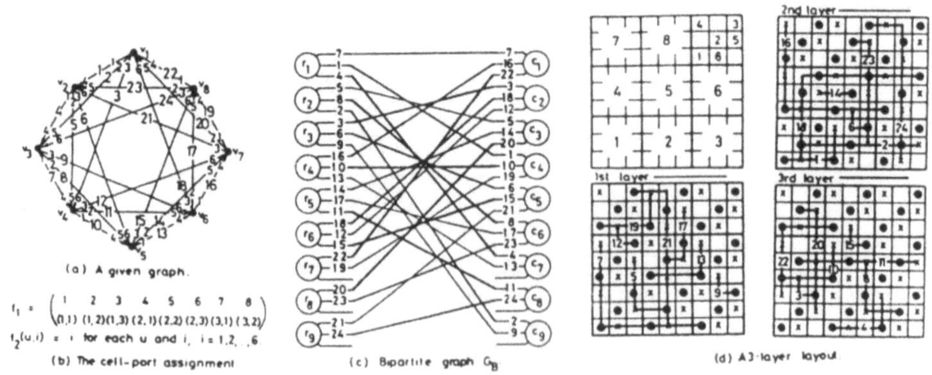

Fig.17. For example 3.

7. Compact Embedding

7.1. Relation Graph

The main problem of the interconnection setting is the layer assignment so that no overlapping arises. We assume that the certain cell-port assignment is defined. We consider the connection patterns which are projected on the space C'_2 in two-dimensional case and on C'_3 in three dimensional case.

Two edges e_1, e_2 of G is the relation R under the cell-port assignment $f = (f_1, f_2)$ if $f(e_1)$ and $f(e_2)$ have no over lapping on C'_2 and C'_3.

Definition 2. Let $G = (V,E)$ be an arbitrary graph. A relation graph $G_R = (W,K)$ constructed from G is a graph such that

i) $W = E$,

ii) $K = \{\{w_i, w_j\};\ w_i\ R\ w_j\}$.

It is easy to observe that the relation graph G_R is of order E and the construction of G_R from G take $O(|E|^2)$ time. Intuitively saying a set of edges in relation R can be realized in one layer with out overlappings.

7.2. Clique Node Cover

A j-clique in a graph G_R is a complete subgraph of G_R with j-nodes. A set of cliques $\{CL_1, \ldots, CL_r\}$ of G_R is call a clique node cover of G_R if

$$W \subset \bigcup_{i=1}^{r} CL_i$$

A minimal clique node cover of G_R is a clique node cover having such minimal r.

Theorem 6. Let G be an arbitrary graph and let r be the cardinality of the minimal clique node cover of G_R. Then G can be realized by CIN with r layers.

From Theorem 6, we may the following algorithm for the optimal layer assignment;

Procedure: Clique node cover
comment: Input $G_R = (W,K)$
begin:
1) Find all the maximal clique of G_R.
2) Find a minimal clique node cover of G_R consisting of maximal cliques.
end.

Proposition 3. The clique node cover procedure is NP-complete.

Therefore it seems impractical to obtain the optimal layer assignment in general. But if we restrict the class of graphs, then the certain optimal layout assignment can be established.

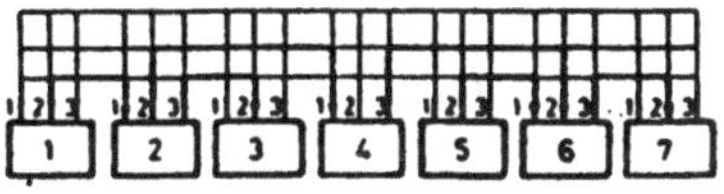

Fig.18. Embed the binary tree in $C_2(7)$.

7.3. Example of the optimal realization

According to the cell-port assignment described in Example 3, we can establish the following embeddings.

(1) 2-D case

It can be easily observed that the overlappings arise only when the edges adjacent with the higher level nodes are realized. Since the level of a n-leaf tree is $\lfloor \log(n-1) \rfloor$, the number of required layers is $\lfloor \log(n-1) \rfloor$. An example of the realization corresponding to Fig.13(b) is illustrated in Fig.18.

(2) 3-D case

According to the cell-port assignment of Fig.14, we obtain the embedding using two layers Fig.19 illustrates the case of height 4 binary tree.

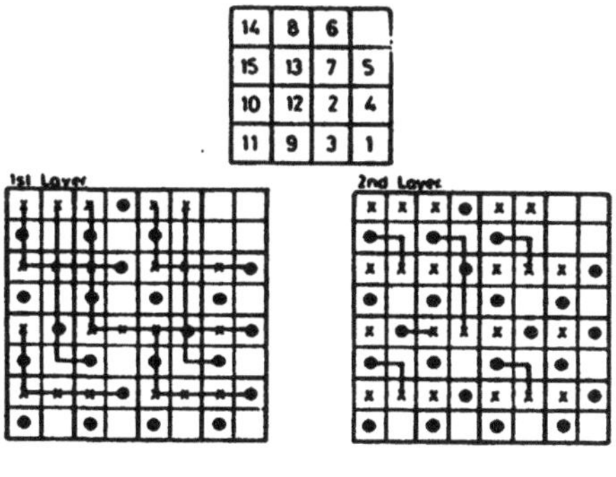

Fig.19. Embed the binary tree in $C_3(4\times4)$.

Proposition 4. An arbitrary n-leaf binary tree can be embedded in 2-D CIN and 3-D CIN using O(nlogn) and O(n) cells respectively.

Conclusion

In this paper we overviewed the architecture of the CECODA system. The user programs written in FORTRAN are firstly transformed to the data flow graphs and if the reduction procedure is applicable further, then we produce the macro data flow graphs. As a such example, we showed the polynomial evaluation problem. In Section 4, 5, 6, the graph embedding algorithms into CECODA are described precisely. This can be considered also to be the processor network construction in CECODA by setting the switches of the CIN according to the input programs. It was demonstrated that this procedure can be performed in polynomial time and an example of binary tree embedding was given. This also indicates that the macro data flow graph of Example 2 can be executed by the processor network of Fig.18 and Fig.19, where the processing unit of this processor network has the ability to compute the program modules of the macro data flow graph.

We described neither the hardware organization or precise control mechanisms of CECODA. The details on these should be referred to other paper[16]. A simulator of CECODA. system has been constructed by restricting the objectives to simple hardware algorithms[17]. In order to construct CECODA, it may also required to study precisely the I/O control mechanism between host computer and CECODA system, the total performance estimation etc.. the other important problem is to exploit the programming language for the CECODA system.

References

[1] J.B.Dennis,"Building Blocks for Data Flow Prototypes," in Proc. Symp. Comput. Arch. 1980.

[2] J.D.Ullman,"Computational Aspect of VLSI," Computer Science Press, 1984.

[3] L.Snyder,"Introduction to the Configurable Highly Parallel Computer," IEEE Computer, pp.47-56, 1982.

[4] S.S.Reddi and E.F.Feustel,"A Restructurable Computer System," IEEE Trans. Comput., Vol.C-27, No.1, pp.1-20, 1978.

[5] P.Lomtong, M.Harao and S.Noguchi,"A Translation Algorithm of FORTRAN IV Language to a Data Flow Graph," 1984 Annual Conference of IECE of JAPAN, 1984.

[6] "------","Data Flow Graph Reduction Algorithm for CECODA System," 1984 Annual Conference of JIPS of JAPAN, 1984.

[7] W.S.Dorn,"Generalization of Horn's Rules for Polynomial Evaluation," IBM J. Res. Develop. 6, 1962.

[8] A.Waksman,"A Permutation Network," JACM, Vol.15, No.1, pp.159-163, 1968.

9 T.Feng,"A Survey of Interconnection Networks," IEEE Computer, pp.12-27, Dec. 1981.

[10] H.J.Siegel,"Interconnection Networks for Parallel and Distributed Processing," Special Issue, IEEE Trans. Comput. Vol.C-30, No.4, 1981.

[11] D.Nassimi and S.Sahni,"A Self-Routing Benes Network," IEEE Trans. Comput. Vol.C-30, pp.332-340, 1981.

[12] L.G.Valiant,"Universality Consideration in VLSI Circuits," IEEE Trans. Comput. Vol.C-30, pp.135-140, 1981.

[13] A.L.Rosenberg,"Three Dimensional VLSI Layout," Math. Sys. Theory, Vol.16, 1983.

[14] M.Harao and P.Lomtong,"A Fast Set-UP and Routing Algorithms for Cellular Interconnection Network," IECE Tech. Commit. of Automaton and Language, July 1984.

[15] H.N.Gabow and O.Karive,"Algorithm for Edge-coloring Bipartitite Graph and Multigraphs," SIAM J. Comput., Vol.11, pp.117-129, 1982.

[16] P.Lomtong, M.Harao and S.Noguchi,"Cellular Interconnection Arrays for Reconfigurable Computing Systems," IECE Tech. Commit. of Automaton and Language, EC83-3, 1983.

[17] "--------","Hardware Algorithm Realization with Cellular Reconfigurable Array," IECE Tech.Commit. of Automaton and Language, AL83-28, 1983.

A SOFTWARE DEVELOPMENT SYSTEM FOR A PROCESSOR ARRAY

A. Graf[1] and R. Pöschel[2]

At the Academy of Sciences, GDR, a parallel processor network for picture processing was designed. One special software development system for this processor array is presented.
This paper consists of three parts. At first, we give a general overview about this software development system. The whole procedure is divided into several steps. Two of them are discussed in more detail in the second and third part of this paper. Thus, in the second part we are concerned with the main working principles of the assembler, the simulator and the synchronization module. Finally, in the third part, the data exchange optimization is considered and one concrete algorithm is developed using a graph-theoretic approach.

There are different possibilities to use parallelism in data processing. The highest degree of parallelism can be achieved by the single instruction-multiple data (SIMD) principle. This principle has been proved as well suited for image preprocessing and statistical analysis of image. This is not true for image pattern recognition. In this case the task of the whole array will be divided into some different subtasks, so we get a multiple instruction-multiple data (MIMD) machine. Programming of such machines is more difficult because of the problems of data exchange and process synchronization.
We have made two restrictions to attach the difficulties with MIMD techniques:

1. The processing time of each subtask does not depend on input data. That means, all program parts must be written in a data independent way.

2. The whole process is divided into two types of operation: internal operations without data exchange and special data exchange operations. It is postulated that all processor elements must process the same type of operation at the same time. The execution time of each subtask processes at the same time does not depend on the concrete operation. Subtasks, which need less time than other subtasks, are filled with dummy operations.

Let us consider an example (Fig. 1):

---

1) Zentralinstitut für Kybernetik und Informationsprozesse der AdW der DDR, DDR-1086 Berlin, Kurstraße 33

2) Institut für Mathematik der AdW der DDR, DDR-1086 Berlin, Mohrenstr. 39

For a specified task three processor elements are used. The first subtask of processor element A requires the time t_{A1}. The first subtask of processor element B requires the time t_{B1}, which is less than t_{A1}. Processor element B will execute dummy operations to fill the time. Processor element C processes its first subtask in the same time as the processor element A, after that time all processor elements exchange their data.
Processor element A requires the time t_{A2} to process its next subtask. Processor element B requires more time to process its next subtask. Processor element A will execute dummy operations. Processor element C requires much more time than the other processor elements to process its next subtask. That's why this subtask is divided into two parts. The first part requires the time t_{B2}. After that time the processor element A and B exchange their data. Processor element C does not need to exchange data. It processes dummy operations. Then the processor elements A and B process their third subtask, and processor element C continues executing the second subtask.

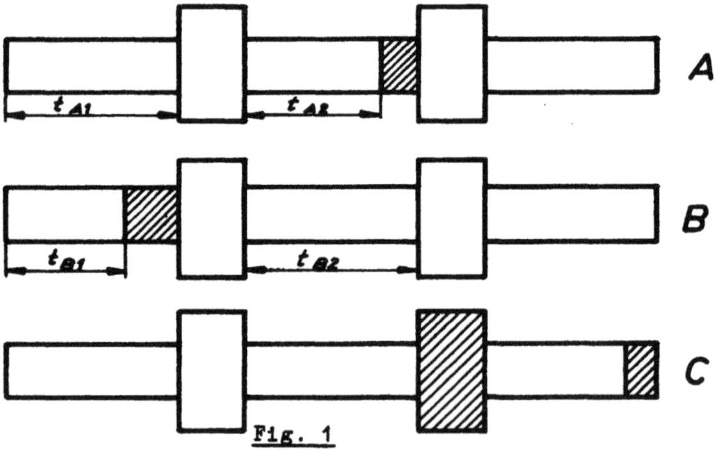

Fig. 1

On the first glance this method seems to be very uneffective because the processor elements perform frequently dummy operations. But the simulation of some tasks has shown, that only some per cent of the whole processing time is needed for data exchange and dummy operations, if the programmer is clever enough and is supported by a simulator showing him the weak (i.e. idle) points of his algorithm. So the lost of efficiency is not high. On the other hand that method allows a high degree of automatisation during software development. All dummy operations may be generated automatically. So we get a tough but very fast software synchronization at program development time. An other effect is the detection of possible run-time errors during program development. Program debugging becomes very easy.

The whole program development procedure is divided into the following steps:

1. Formulation of the problem
2. Algorithm development, division into subtasks
3. Dislocation of the subtasks
4. Data exchange optimization
5. Detailed programming of the internal operations
6. Assembly, simulation and automatical synchronisation
7. Evaluation of the simulation results and correction.

Two software development tools have been developed until now. The first program optimizes data exchange between the processor elements. The second program performs assembly, simulation and synchronization.
The programs are written for running on any PDP-11 compatible computer. The processor element was assumed to be the wellknown one-chip microcomputer Z8.

In the following, we are going to describe the working principles of the assembly- and synchronization program.
The main task of this module is to assemble the source program and to insert needed dummy operations to get the software synchronization.
The input is a line-structured editing file. The program puts out a disk file containing the duplicate of the whole processor array and a list file.
After determining the hardware system to be simulated the assembler code of the first subtasks for each processor element is read in, assembled and inserted into the duplicate file. After that, the program execution is simulated for all processor elements. The simulation begins at the starting point determined by the hardware to be simulated. The initial time is set to 0, the time units counted are ticks of the hardware timer. One of the simulation results is exact information about the program execution time for each processor element. This set of time information is used by the synchronization algorithm.
The synchronization is not trivial, because there is only a limited set of dummy operations. It may occure, that all subtasks have to be extended to make it possible to fill the time with dummy operations from the given instruction set.

Let us show an example (Fig. 2):
Let there be a system containing only two active processor elements. Processor element A processes its first subtask during the time t_{A1}. Processor element B processes its first subtask 4 ticks faster. The shortest dummy operation requires six ticks, the next 10. The shortest time, on which the processor elements can be synchronized, is $t_{A1}+6$ ticks. Processor element A will execute a 6-ticks dummy operation. Processor element B a 10-tick one.

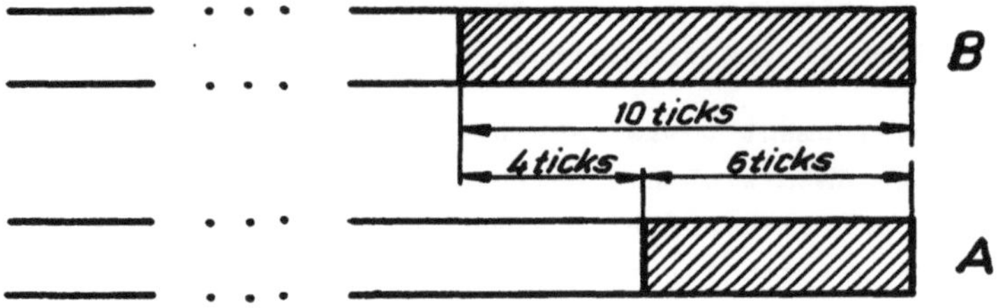

Fig. 2

The synchronization algorithm is divided into two parts to get a good relation between the size and the execution speed of the development program. During the first part the whole time for the current subtasks, including dummy operations, and all time differences to be filled with dummy operations are computed. If at least one of the time differences is too large, a message is generated. The programmer has to correct his algorithm or to use other ressources.

During the second part the dummy operations are generated and inserted into the duplicate and list files. A lookup-table is used to find the optimal combination of dummy operations for each possible time difference. After the synchronization is completed, the program execution of one processor element is simulated to get exact data about the idle intervalls each processor element.

The code necessary for data exchange is pre-generated during the automatical data exchange optimization step, which is described below. It is synchronized similar to the subtask code.

Now, let us consider another step, namely the data exchange optimization. For data exchange several connections between the processor elements must be realized; we say that a given communication structure must be realized in the array. This can be done in different ways, and the problem to find the optimal variant is called here the communication problem.

It turns out that, from the mathematical point of view, the communication problem is equivalent to a graph-theoretic colouring problem (for all graph-theoretic notions and results used in the following we refer e.g. to /1/).

In order to simplify the presentation we shall demonstrate this by considering only one column (or row) of the processor array.

Let PE_0, PE_1, ..., PE_{n-1} (for short 0,1,...,n-1, n natural number) be the processors which are connected in one column as shown in Fig. 3.

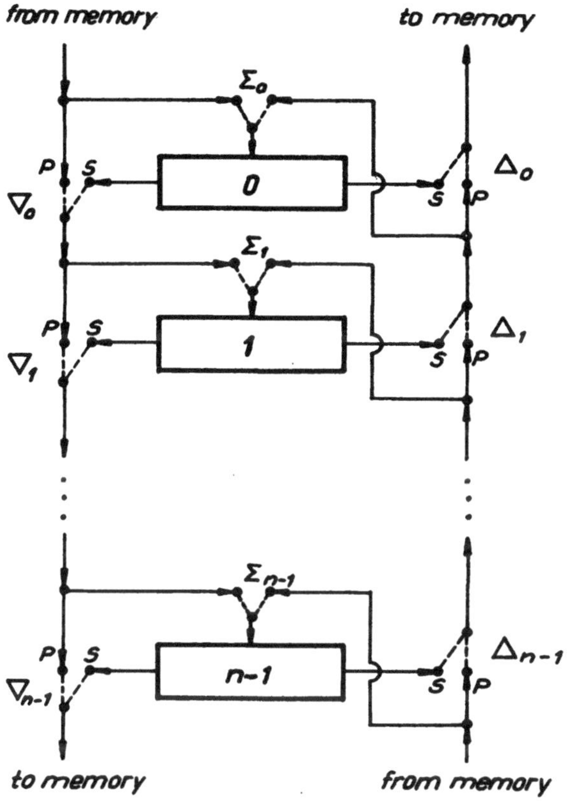

Fig. 3

The processor PE_i controls the switches ∇_i, Δ_i and Σ_i ($i=0,1,\ldots,n-1$) what enables it to send (switches ∇ or Δ in state "S"), to let pass (downwards or upwards; switches ∇ or Δ in state "P") or to receive (from above or from below; switches Σ in state "1" or "0") information signals on the common connection wires.

A configuration is the set of all states (positions) of the switches in a given moment.

Now, a communication structure π can be described by a set of pairs (a,b), where $a,b \in \{0,1,\ldots,n-1\}$.

Then $(a,b) \in \pi$ means that there has to be a connection from processor PE_a to PE_b.

Example: Fig. 4 represents the following communication structure ($n=6$):

$$\pi = \{\underbrace{(0,5),(1,4),(2,3)}_{\text{"downwards"}},\ \underbrace{(5,2),(3,1),(3,0)}_{\text{"upwards"}}\}$$

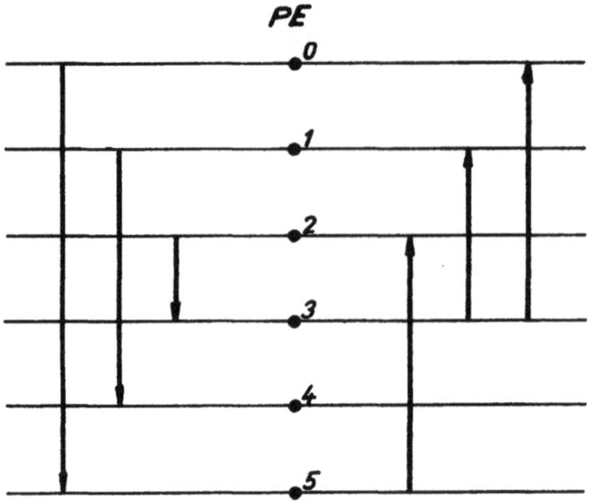

Fig. 4

We say, a configuration K realizes a communication structure π if the configurations realizes every connection $(a,b) \in \pi$ such that no connection between the processors is used for more than one pair $(a,b) \in \pi$ (this condition enables conflict-free information flow).
More general, a set of configurations $K_1, K_2, \ldots, K_t$ realizes a communication structure π, if there exist communication structures $\pi_1, \ldots, \pi_t$ such that

$$\pi \subseteq \pi_1 \cup \pi_2 \cup \ldots \cup \pi_t$$

and K_i realizes π_i $(i=1,\ldots,t)$, i.e., π can be realized in t steps by using the configurations $K_1, \ldots, K_t$ consecutively.
Let $t_{min}(\pi)$ be the minimal t such that there exist $K_1, \ldots, K_t$ which realize π.
We restrict our considerations to communication structures π for which every processor receives from at most one other processor, i.e. it holds

$$(a,b),\ (a',b) \in \pi \implies a=a'$$

(such communication structures are called injective).

Now we are able to give a precise formulation of the communication problem:

For a given injective communication structure find $t_{min}(\pi)$ and a set $K_1, \ldots, K_{t_{min}(\pi)}$ of configurations realizing π.

We are going to give an optimal solution to this problem.

For a given communication structure π, divide π into two reduced parts:

$$\pi^{\nabla} = \{(a_i,b_i) \mid (a_i,b_i) \in \pi \ \& \ a_i < b_i \ \& \ b_i = \max\{b \mid (a_i,b) \in \pi\}\}$$

$$\pi^{\Delta} = \{(a_i,b_i) \mid (a_i,b_i) \in \pi \ \& \ a_i > b_i \ \& \ b_i = \min\{b \mid (a_i,b) \in \pi\}\}$$

Then, in general, $\pi \supsetneq \pi^{\nabla} \cup \pi^{\Delta}$
but without difficulties one can show that

$$t_{min}(\pi) = \max\{t_{min}(\pi^{\nabla}),\ t_{min}(\pi^{\Delta})\}.$$

Interpret the elements of π^{∇} and π^{Δ} as intervals. Then it is easy to see that two elements (a,b), (a',b') of π^{∇} can be realized by o n e configuration if and only if the intervals do not overlap, i.e., iff $b \leq a'$ or $b' \leq a$, (analogously for π^{Δ}). From the graph-theoretic point of view this means that the <u>chromatic number</u> $\chi(G(\pi^{\nabla}))$ of the corresponding interval graph $G(\pi^{\nabla})$ is an upper bound for $t_{min}(\pi^{\nabla})$ (the vertices of $G(\pi^{\nabla})$ are the elements of π^{∇} , the edges are all pairs of vertices which (as intervals) do overlap).
On the other hand, the so-called <u>clique number</u> $\omega(G(\pi^{\nabla}))$ of the interval graph $G(\pi^{\nabla})$ is a lower bound for $t_{min}(\pi^{\nabla})$ in fact, $\omega(G(\pi^{\nabla}))$ equals the greatest number of mutually overlapping intervals, therefore no both of which can be realized by the same configuration).

Thus we have

$$\omega(G(\pi^{\nabla})) \leq t_{min}(\pi^{\nabla}) \leq \chi(G(\pi^{\nabla})).$$

It is well-known (cf. e.g. /1/) that interval graphs are perfect graphs, i.e. their clique number and chromatic number coincide. This leads to the following theorem:

<u>Theorem:</u> Let π be an injective communication structure (over a processor column). Then

$$t_{min}(\pi) = \max\{t_{min}(\pi^{\nabla}),\ t_{min}(\pi^{\Delta})\},$$

where

$$t_{min}(\pi^{\nabla}) = \omega(G(\pi^{\nabla})) \text{ and, analogously,}$$

$$t_{min}(\pi^{\Delta}) = \omega(G(\pi^{\Delta})).$$

It remains the task to find explicitely the configurations which realize π in $t_{min}(\pi)$ steps. Having a colouring of $G(\pi^{\nabla})$ and $G(\pi^{\Delta})$ at hand, say with colours 1,2,...,t ($t = t_{min}(\pi)$), let

$$\pi_i = \left\{(a,b) \;\middle|\; \begin{array}{ll} \text{There is } (a,b') \in \pi^{\nabla} & \text{with colour } i, \text{ if } a < b; \\ \text{There is } (a,b') \in \pi^{\Delta} & \text{with colour } i, \text{ if } a > b \end{array}\right\},$$

(i=1,2,...,t). Then $\pi = \pi_1 \cup \ldots \cup \pi_t$ and, for every i=1,...,t, there exists one configuration, say K_i, which realizes π_i (by construction there can be no conflict giving the switches ∇, Δ, Σ the positions necessary for realizing π_i).

In the case under consideration (one column only), there is a very easy algorithm to colour $G(\pi^{\nabla})$ with $\omega(G(\pi^{\nabla}))$ colours (analogously for $G(\pi^{\Delta})$):
Run trough the vertices of $G(\pi^{\nabla})$ (one times only!) in the order that the first components of the vertices (= intervals) increase, and assign to every interval the least possible colour which does not conflict with the already coloured intervals.

In the example given above (Fig. 4), 3 colours are needed (because $\omega(G(\pi^{\nabla})) = 3$, 3 overlapping intervals!). The mentioned algorithm provides π_1 (colour 1), π_2 (colour 2) and π_3 (colour 3) as shown in Fig. 5.

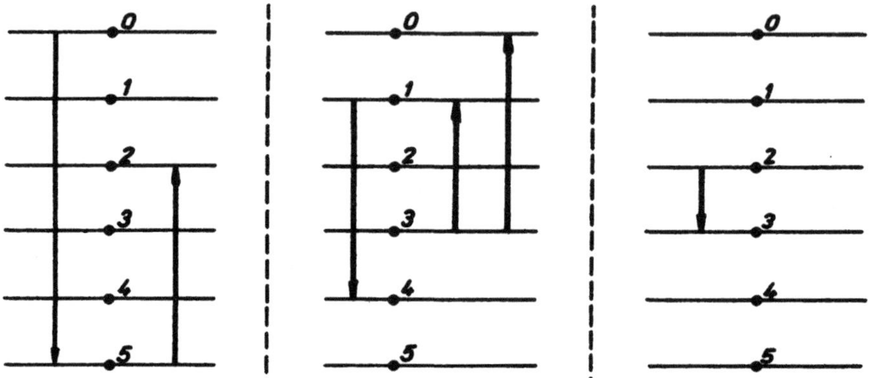

$\pi_1 = \{(0,5),(5,2)\}$ | $\pi_2 = \{(1,4),(3,1),(3,0)\}$ | $\pi_3 = \{(2,3)\}$

Fig. 5

The corresponding configurations K_1, K_2, K_3 which realize π_1, π_2, π_3, resp., and, moreover, also $\pi = \pi_1 \cup \pi_2 \cup \pi_3$ are given in the following table (all position of switches which are not mentioned, can be arbitrary).

switches in position	S	P	1	0
Configuration K_1	∇_0, Δ_5	$\nabla_1, \nabla_2, \nabla_3, \nabla_4, \Delta_3, \Delta_4$	Σ_5	Σ_2
K_2	∇_1, ∇_3	$\nabla_2, \nabla_3, \Delta_1, \Delta_2$	Σ_4	Σ_0, Σ_1
K_3	∇_2		Σ_3	

REFERENCES

/1/ M.Ch. Golumbic, Algorithmic Graph Theory and Perfect Graphs. Academic Press, New York, London, Toronto 1980.

VLSI SYSTOLIC ARRAY FOR THE FAST SOLUTION OF SYSTEM OF LINEAR EQUATIONS

Jozef Miklośko [1]

1. Introduction

Let us have the linear algebraic equation system

$$AX = B \qquad (1)$$

where $A=(a_{ij})$ is a regular matrix of n-th order, $X=(x_i^{(j)})$ and $B=(a_{i,n+j})$ are n/m matrices, $i=1,2,\ldots,n$; $j=1,2,\ldots,m$.

Solve (1) by Gauss-Jordan elimination with partial pivoting [1]. Let us construct the matrix $C=(A|B)=(a_{ij})$, $i=1,2,\ldots,n$; $j=1,2,\ldots,n+m$. In this method after appropriate exchange of rows, the matrix A is eliminated by the column to the unit matrix I. The same operations are done on B, so that from C we get (I|X), i.e., the solution X is obtained after n-th step of elimination in the place of the B.

The Gauss-Jordan elimination can be described by these steps [1]:

Let $a_{ij} = a_{ij}^{(o)}$.

For $s=1,2,\ldots,n$ do:

1) search such i that $a_{is} = \max_{j\geq s} |a_{js}^{(s-1)}|$,
2) exchange s-th and i-th rows of the matrix,
3) calculate $c_s=1/a_{ss}^{(s-1)}$,
4) multiplicate s-th row of the matrix with c_s,
5) for $i=1,2,\ldots,n$ do:

   for $j=s,s+1,\ldots,n+m$ do:

   $a_{ij}^{(s)} := a_{ij}^{(s-1)} - a_{is}^{(s-1)}(a_{sj}^{(s-1)}/a_{ss}^{(s-1)})$

   where the expressions in the bracket are calculated in the step 4).

From C we get (I|X) where $a_{i,n+j}^{(n-1)}=x_i^{(j)}$. The value of $d=\det|A|$ can be calculated too:

$$d=(-1)^J \prod_{i=1}^{n} a_{ii}^{(i-1)} \qquad (2)$$

where J is the total number of exchanges of rows, i.e., after each exchange in the 2-nd step of algorithm the sign of the intermediate product must be changed.

Gauss-Jordan-Rutishauser Algorithm (GJR) [1] makes after each step of elimination a cyclic shift of columns to the left and rows upwards, so that we always get the pivot into the position a_{11}. Only parallel calculation of this algorithm can take all advantages of this fact.

In this paper we suggest specialized VLSI systolic array SASLE (Systolic Array for System of Linear Equations) for the calculation of (1) and (2) by modification of GJR algorithm in which matrix C is shifted to the left and cyclically upwards after each elimination's step (GJR1). We describe the scheme of a given computer, its specialized set of

1) Institute of Technical Cybernetics, Slovak Academy of Sciences, 842 37 Bratislava, Dúbravská 9, Czechoslovakia.

instructions and the program executing competent algorithm for matrix C of n/(n+m) order. Finally, we also briefly evaluate our computer from point of view its program's complexity.

Input data of (1) and (2) are integers n,m,b, boolean p and real matrix C. Output is determinant d and matrix X.

Integer b enables us to have variable-precision arithmetic in our computer. It determines how many bits in registers and in one memory cell will be reserved for storing of floating point number and, for setting of b-bits, floating point arithmetic operations -, × and /. We suppose that after input of b, hardware in our computer is automatically set to b-bits arithmetic, which enable us to achieve sufficient precision of results. Boolean p determines necessity of partial pivoting. If p=1, then partial pivoting is required; if p=0, then algorithm can be executed without pivoting (e.g. this is the case if A is a positive definite or a diagonally dominant matrix).

2. VLSI systolic array SASLE

2.1 Introduction

The first systolic system for the calculation (1) was described in [2]. The system implements the LU decomposition of the matrix A on a hexadiagonal array of inner product step processors with one built-in processor for calculating the reciprocal value. Specialized systolic arrays for (1) and for matrix inversion with GJR without pivoting are in [3]. The universal almost homogeneous systolic array which can solve on one array of processors by GJR without pivoting some (or all) of these problems: (1), (2) and A^{-1} are in [4]. In the next part we describe VLSI systolic array SASLE, which is specialized for GJR1 for solution of (1) and (2).

2.2 Structure of SASLE

SASLE will be composed of the array of n.(n+m) processors p_{ij} (i=1,2,...,n; j=1,2,...,n+m) with n rows and n+m columns which will be interconnected by the square communication network. Our systolic array SASLE has, except for p_{11}, a homogeneous structure given in the Fig.1.

The processor p_{11} executes the function of control unit, searches a pivot, calculates its negative reciprocal value and determinant d by the multiplication of pivots. It has 8 indexregisters (rn_{11},rm_{11},rb_{11},rp_{11}, rs_{11},ri_{11},rj_{11},rc_{11}) and 4 floating point (FP) registers (Ra_{11},Rx_{11},Ry_{11}, Rd_{11}), on the contents of which FP multiplication and negative reciprocal value is calculated. All other processor p_{ij}, $i\neq 1 \wedge j\neq 1$ are the same. They have only 3 FP registers (Ra_{ij},Rx_{ij},Ry_{ij}) on the contents of which the FP operation $Ra_{ij}:=Ra_{ij}-Rx_{ij}\times Ry_{ij}$ is executed.

The communication network is rectangular of the type "each with four neighbours", moreover the first row of processors is cyclically connected with all other rows in the array.

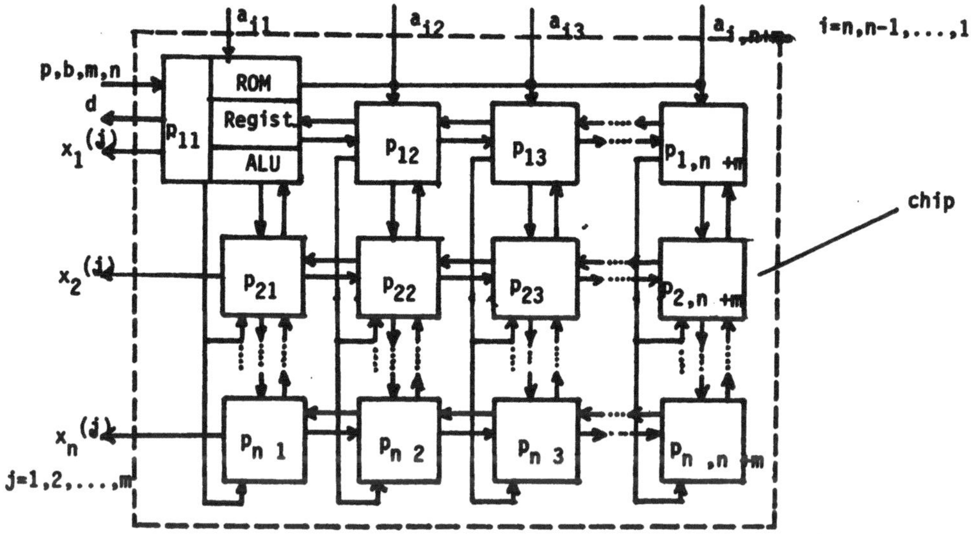

Fig.1.

2.3 Instructions and program for SASLE

The following symbols have been chosen to describe our systolic program:

$\forall i$ means: $i=1,2,\dots,n$,

$\forall j$: $j=1,2,\dots,n+m$,

$\forall i(k)$: $i=1,2,\dots,k-1,k+1,\dots,n$,

$\forall i(i \leq n-s+1)$: $i=1,2,\dots,n-s+1$,

$\forall j(j \leq m)$: $j=1,2,\dots,m$,

whereby $i, j, n, n+m, s$ are stored in the indexregisters $ri_{11}, rj_{11}, rn_{11}, rm_{11}, rs_{11}$. $\Downarrow, \Uparrow, \Rightarrow, \Leftarrow$ means direction of data flow in the given operation. Prefix par- (p-in the description) means its parallel execution. Symbol $f \wedge g$ means parallel realization of the instruction f and g whereby the continuation of program can be done only if both f and g are finished.

The instructions executed by p_{11} are in Table 1. The instructions calculated by p_{11} and some other processor p_{ij} (only by some other processors p_{ij} except p_{11}) are in the Table 2 (Table 3). Where necessary, a brief description of the activity of each processor p_{ij} is given in the bracket.

Table 1

No.	Instruction	Description
1	parinput $\Rightarrow (n \wedge m \wedge b \wedge p)$;	p-input of n,n+m,b,p into $rn_{11}, rm_{11}, rb_{11}, rp_{11}$. Set hardware of all necessary instructions on manipulating with b-bits FP words.
2	output $\Leftarrow Rd_{11}$;	Output of d to the left.
3	$ira_{11} := irb_{11} + k$;	Integer add. ira_{11} and irb_{11} are indexregisters, k is integer.
4	if CON then L;	If CON is true, then jump on integer label L.
5	$Rx_{11} := Ra_{11}$;	Restoring of Rx_{11} by Ra_{11}.
6	$Rd_{11} := Rd_{11} * Ra_{11}$;	FP multiplication.
7	$Ry_{11} := -1/Ry_{11}$;	FP negative reciprocal value.

Table 2

8	ZERO;	Reset all registers and indexregisters in SASLE. Set $Rd_{11} := 1$.
9	for $\forall i \forall j$ parinput $a_{ij} \Downarrow Ra_{ij}$;	p-input of a_{ij} from the top to the bottom into Ra_{ij} (for $\forall i \forall j$: take $Ra_{i-1,j}$, store into Ra_{ij} and sent it into $Ra_{i+1,j}$ till a_{nj} is not in Ra_{nj}; Ra_{oj} means input from outside).
10	for $\forall i$ partransfer $Ra_{i1} \Rightarrow Rx_{ij} \forall j(1)$;	p-transfer of the Ra_{i1} to the right into Rx_{ij} $\forall j(1)$ (for $\forall i$: take Ra_{i1}, store it into Rx_{i2}; for $\forall i \forall j (j \neq 1 \wedge j \leq n+m)$: take Rx_{ij} and send it into $Rx_{i,j+1}$, until this wave does not come to $Rx_{i,n+m}$).
10a	partransfer $Ry_{11} \Rightarrow Ry_{1j} \forall j(1)$;	Analogical to previous instruction, but only in the first row of processors.
11	for $\forall i \forall j$ parshift $Ra_{i,j-1} \Leftarrow Ra_{ij}$;	p-shift of the Ra registers to the left, whereby Ra_{i1} $\forall i$ are canceled (for $\forall i \forall j(1)$: take Ra_{ij} and store it into $Ra_{i,j-1}$).
12	for $\forall i \forall j$ parcyclshift (a,b) $Ra_{ij} \Uparrow Ra_{i-1,j}$;	In the first b rows of processors a-times p-cyclic shift of the Ra registers upwards (for $\forall i (i \leq b) \forall j$ a-times do: take Ra_{ij} and store it into $Ra_{i-1,j}$, whereby $Ra_{oj} \equiv Ra_{bj}$).
12a	for $\forall i (i \leq n-s+1)$ par cyclshift$(1, rc_{11})$ $Ra_{i1} \Uparrow Ra_{i-1,1}$;	Analogical to previous instruction but only in the first rc_{11} processors of the first column of array.
13	for $\forall i \forall j (j \leq m)$ paroutput $\Leftarrow Ra_{ij}$;	p-output of X, i.e., Ra_{ij} $\forall i \forall j (j \leq m)$ (for $\forall i$ m-times do: p-output Ra_{i1} to the left, for $\forall j$ $(j \leq m)$ take Ra_{ij} and store it into $Ra_{i,j-1}$).

Table 3

14	for ∀i(1) ∀j(1) pardo $Ra_{ij}:=Ra_{ij}-Rx_{ij}\times Ry_{ij}$;	p-performance of elimination step on the appropriate registers and storage of results into Ra_{ij}.
14a	for ∀j(1) pardo $Ra_{1j}:=Ra_{1j}-Rx_{1j}\times Ry_{1j}$;	Analogical to previous instruction, but only in the first row of processors.
14b	for ∀j(1) pardo $Ra_{1j}:=0$;	Reset of Ra_{1j}, ∀j(1).

The program P for SASLE is the following (the first column gives the number of the given step; the second gives the number of active processors in this step; if the step is executed only by p_{11} then the second column is blind):

```
0   n(n+m)          ZERO;
1                   parinput⇒(n∧m∧b∧p)∧
    n(n+m)          for ∀i ∀j parinput a_ij⇓Ra_ij;
2                   rs_11:=rs_11+1∧if rp_11=0 then 11;
3                   Rx_11:=Ra_11∧ri_11:=0∧rj_11:=0∧rc_11:=rn_11-rs_11+1;
4                   if rs_11=rn_11 then 11;
5                   rj_11:=rj_11+1∧
    n-s+1           for ∀i(i≤n-s+1) parcyclshift(1,rc_11) Ra_i1⇑Ra_i-1,1;
6                   if abs(Ra_11)≤abs(Rx_11) then 8;
7                   Rx_11:=Ra_11∧ri_11:=rj_11;
8                   if rj_11<rc_11 then 5;
9                   if ri_11=0 then 11;
10  (n-s+1)(n+m)    for ∀i(i≤n-s+1)∀j parcyclshift(ri_11,rc_11) Ra_ij ⇑
                    Ra_i-1,j;
11  n(n+m)          for∀i partransfer Ra_i1⇒Rx_ij ∀j(1);
12                  Ry_11:=-1/Ra_11∧Rd_11:=Rd_11 * Ra_11;
13  n+m             partransfer Ry_11⇒Ry_1j ∀j(1)∧
    n+m-1           for ∀j(1) pardo Ra_1j:=0;
14  n+m-1           for ∀j(1) pardo Ra_1j:=Ra_1j-Rx_1j * Ry_1j;
15  n(n+m-1)        for ∀j(1) partransfer Ra_1j⇓Ry_ij ∀i(1);
16  n(n+m)-2n-m+1   for ∀i(1) ∀j(1) pardo Ra_ij:=Ra_ij-Rx_ij * Ry_ij;
17  n(n+m-1)        for ∀i ∀j parshift Ra_i,j-1⇐Ra_ij;
18  n(n+m)          for ∀i ∀j parcyclshift(1,rn_11) Ra_ij⇑Ra_i-1,j;
19                  if rs_11<rn_11 then 2;
20                  output⇐Rd_11∧
    n               for ∀i ∀j(j≤m) paroutput⇐Ra_ij;
```

Program P is distributed through the array of processors in such a manner that in each processor only the number of step and microprogram of competent instruction is stored. p_{11} synchronizes the calculation SASLE, sets the step number just being executed, and sends this information to all processors in the array. Each processor obtains this information; if it is devoted to it, then it calculates the competent instruction and after finishing sends the signal back to p_{11}. If p_{11} obtains from active processors the necessary number of signals given in the second column of program P, then the next step of the program can be executed.

2.4 Complexity of the program P

The number of instructions executed in P are given in Table 4 (instructions No. 1,2,5,8,9,13 are not considered, No.6 and 14b are not counted because of their parallel calculation with other instructions).

Table 4

Instruction No.	3	4	7	10+10a	11	12+12a	14+14a
Their number	2n	n^2+5n	n	$2n^2-n$	n	n^2+n	2n
Pivoting only	-	n^2+n	-	-	-	n^2	-

We see that in P arithmetic takes only O(n) steps, but both transfers and pivoting take $O(n^2)$ steps. Table 4 was constructed under these assumptions:

a) in pivoting we always suppose the worst case, i.e., an exchange of rows with maximal distance,

b) I/O instructions are not considered and some simple restoring index-registers and registers instructions are neglected,

c) in one row of program we evaluate only the longest instruction (we suppose that instruction No.4 (No.7) takes more time than No.3 (No.6)).

3. Remarks

a) We suppose that by hardware realization of SASLE high performance machines can be created, which can calculate with high speed and accuracy large-scale dense systems of linear algebraical equations. The real cost and performance of SASLE must be of course proved at its simulation or practical realizartion, which was not our goal.

b) In our computer there are not tests for calculation of residua in (1), i.e., AX-B. If we want to have this test in program P, then P must be slightly modified, since A and B are canceled in the course of calculation.

c) Our computer can also be modified in such a manner, that together with (1) and (2) it will also calculate matrix inversion A^{-1} by algorithm GJR with partial pivoting. Some specialized systolic arrays for (1), (2) and A^{-1}, calculated by GJR without pivoting, are described in [4].

Acknowledgement

This work was supported by the Alexander von Humboldt Foundation during the author's stay at the Institut für Informatik, Technical University Munich in the summer of 1984. I cordially thank Prof.Dr.F.L.Bauer for help and criticism of this paper.

References

[1] Bauer, F.L., Heinhold, J., Samelson, K., Sauer, R.: Moderne Rechenanlagen, B.G.Teubner Verlag, Stuttgart, 1965.

[2] Kung, H.T., Leiserson, C.E.: Systolic arrays for VLSI, Proc. Sparse Matrix 1978, SIAM, 1979, 256-282.

[3] Mikloško, J.: Systolic system for the linear equation system and matrix inversion, Comp.and Artif.Intel., 2, 1983, 4, 361-372.

[4] Mikloško, J., Zaťko, B.: Universal systolic array processors for fast matrix operations, Proc.Artif.Intel.Inf.-Contr.Syst.of Robots, I. Plander (editor), Elsevier Science Publ.B.V. (North Holland), 1984, 259-263.

SESSION: CELLULAR SYSTEMS

CELLULAR SYSTEMS

Tamás Legendi[1/]

Abstract

After explaining the trend towards highly parallel working, homogeneous computing systems a short overview of a cellprocessor project is given. Mainly the results published in this Proceedings are summarized.

Key words: Systolic array, instruction broadcasting, array processor, cellular processor, cellular algorithm, cellular program, VLSI algorithm, WSI, cellware, Megacell machine.

1. Introduction

In the past decade /V/LSI made possible, user requirements forced R&D of parallel processors. Systems with regular, homogeneous structure are well-suited for microelectronic and electronic implementation, production and testing and they are suitable for some classes of parallel processing.

There exists a very natural microelectronic trend for different level homogeneous components as memories, /F/PLA-s, PAL-s, gate-arrays, ULA-s. Different classes of partially homogeneous design exist as iterative array, cellular array concept; PROM based, multiplexer based design methodology. These homogeneous components/methodologies have parallel processing not as a primary goal but may be applied in parallel systems, too.

Many results are achieved in the direction of parallel, partially homogeneous systems - array processors, data-flow machines, parallel and pipe-lined arithmetics, multi-microprocessor systems [7], [27]. Here we concentrate on systems based on relatively small /1-5 % of an 8 bit microprocessor/ repetitive elements. In our cellprocessor systems they will be called microcells in contrast to bigger repetitive elements called macrocells.

The best-known, wide range results in this direction are that of Kung [17-20] and his followers [8] in the field of systolic arrays. A lot of papers have been published on VLSI algorithms. First special purpose chips have been designed then the interest has turned towards programmable structures /PSC/. A very remarkable experiment is the blue

1/ Research Group on Theory of Automata Hungarian Academy of Sciences H-6720 SZEGED, Somogyi u. 7.

chip project of Snyder [28], [29].

The author has proposed in 1976-77 a microcell based fully homogeneous, highly parallel working cellprocessor system [24] and its programming methodology [21]. Based on this proposal a project has been started, the ideas were continuously being refined according to the technological limitations, growing cellprogramming experience and the designed sophisticated cells.

2. Mathematical results

As the project started on the basis of the homogeneous infinite cellular automata notion some results were needed for inhomogeneous finite cellular spaces [21].

3. Simulation systems

Different large-scale simulation systems have been designed and implemented for testing cellular algorithms [2] [23] [1] cellular microprograms and programs [4].

4. Cellular algorithms

A methodology has been worked out [21] and refined [25], [13] hundreds of algorithms have been designed and tested [21], [9], [2], [3], [10], [12], [14], [32].

A cellular algorithm may operate efficiently on an array of data. They are processed in a bit-parallel and bit pipe-lined fashion. The data move continuously through processing elements /groups of cells executing a well-defined operation/. Processing elements are software defined, the cellprogrammer handles an inhomogeneous cellular space. Cellprocessing methodology changed and developed much according to the growing experience and the designed concrete cells.

5. Cellprocessor architecture

A cellprocessor may be a universal or a special-purpose one. Although we have designed special purpose cellprocessors, too /an LSI based sorter, and traditional component based convolution processors for picture processing/, here we concentrate on universal cellprocessors.

The most important part of them is the

5.1 Cellular field

It consists from identical cells. /Full hardware homogeneity./ By loading /internal states in the case of LSI implementations or transition functions in the case of traditional component implementation/ or masking /in the case of our cellprocessor model/ the cells may be assigned different transition functions /microprograms/. Software inhomogeneity and its efficient implementation are ensured in this way.

The cellular field emulates a fully connected cellular space with the extra possibility of immediate teletransporting of byte streams. This feature makes much more easier the mapping of complex cellgorithms into the cellular field.

The control of the cellprocessor is concentrated in the

5.2 Cellular control processing unit

The CCPU is a programmable device in the new LSI based systems that coordinates the loading of the full cellprocessor, transition function/microprogram execution and I/O data streams.

In the traditional component based systems CCPU looks like as an extended microcontroller.

The continouos flow of data to be processed between the host and the cellular field is directly programmed in

5.3 The I/O processor

which interfaces the host processor data transfer capabilities with the processing capabilities of the cellular field. Using queues in a relatively big I/O buffer and executing some format conversions the I/O processor /pair/ moves data from and to the host memory - I/O buffer - cellular field edge buffer.

More detailed descriptions of the new LSI based architectures may be found in [26], [31], on the cellprocessor model in [30] and on traditional component based systems in [33].

In the existing terminology a microprogrammable cellprocessor may be called as a programmable systolic array with microinstruction broadcasting.

6. Cellular programming and microprogramming

A cellular algorithm gives a frame, a specification when we must /micro/program a concrete cellular architecture, a concrete cellware.

In the LSI based systems transition functions should be microprogrammed e.g. transformed into a sequence of microinstructions that drive directly in a serial manner the cells.
Time-space inhomogeneity /PE assignment and execution/ is ensured through internal state loading and CCPU programs.
The flow of data and some format conversions are under the control of I/O programs.
These three different cooperating components may be integrated into one language processor. Microprogram optimization algorithms make easier the coding, more efficient the run.
The process of making code from cellular algorithms is explained and illustrated in [15].

7. Host software

The simplest first approach is to ensure cellular operators /macros, new instructions/ for the end-user in his own usual programming environment. Most probably only this simple solution may spread fast and wide enough. The elements for such a system /library of operators, language processor extension, handler/ were designed and tested for the cellprocessor model [4]. Such a system seems to be moderate in size and complexity. After some experience users may become interested to define their own operators. Higher level cell-/micro/program languages may satisfy this need. It seems to be not very probable that full tasks will be coded in a cellular language rather only those parts that need high speed execution and may really be cellprogrammed while the remainder in the usual source languages will be coded.

8. Cellprocessor model

A 256 two-state microprogrammable cellprocessor / 2 TTL chips/cell, 1000 chips in the full system / has been built to show the principle operating and to learn lessons before the development of the Megacell machine.
As here the cells are simplified the cellular field /half of the system in size/ raised no problems. CCPU was transparent, the I/O instruction set proved to be problematic.

It was hard to cellprogram such a small space.
All cellprograms are written now directly in a microassembler. /Higher level cellular language and its processor is under design based on this experience and on the need of more complex cellgorithms./
The host-software is nearly as complex as it will be on a real machine so its implementation was an important step towards the final system.

As the two-state integrated chip contains extended, more sophisticated cells which allow shorter microprograms for the same function at the same microinstruction execution speed /N-MOS/ while one 28 pin chip replaces more then 200 TTL chips - the size and the computing power of the Megacell machine may be estimated now more precisely.

9. Applications

A lot of efforts were made to select appropriate application areas and to assemble application case studies.
Picture processing seems to be the most fruitful area where special hardware for preprocessing has been designed, dozens of cellgorithms [15] , [32] and cellprograms for different architectures have been worked out and tested.
The design of a macrocell based finite element machine and finite element cellgorithms are going on[6].
Digital signal processing and filtering proved also to be a very adequate application area.
Process control case studies including tested cellgorithms show the possibility of cellprocessing in this area.
The simulation of large-scale systems including stochastic simulation is available by cellprogramming.
The solution of partial differential equations and systems of linear equations is natural.

The most general use is any kind of array processing that can be cellprogrammed and efficiently called from the host.

References

[1] Almási, J., Medvey, M.: The CELLAS 2.1. Cellular Simulation System. Presented at II. Int. Workshop on Parallel Processing by Cellular Automata and Arrays, Berlin, 1984.

[2] Bara, T.: Division in O/n/ Steps in Cellular Space. Presented at II. Int. Workshop on Parallel Processing by Cellular Automata and Arrays, Berlin, 1984.

[3] Bara, T.: The realization of tree Algorithms on a Cellprocessor. Presented at II. Int. Workshop on Parallel Processing by Cellular Automata and Arrays, Berlin, 1984.

[4] Barsi, I.: Cellular Processor Model Host Software Architecture. Presented at II. Int. Workshop on Parallel Processing by Cellular Automata and Arrays, Berlin, 1984.

[5] Batcher, K. E.: Design of a Massively Parallel Processor. IEEE Transaction on Computers, C-29, 9, 1980, pp. 836-840.

[6] Bernáth, L.: Cellular Finite-Element Algorithm. Presented at II. Int. Workshop on Parallel Processing by Cellular Automata and Arrays, Berlin, 1984.

[7] Bernutat-Buchman, U., Rudolph, D., Schlosser, K.-H. /eds/: Parallel Computing 1. Eine Bibliographie /5161 references/ 2. Auflage Bochumer Schriften zur Parallelen Datenverarbeitung 1, Rechenzentrum der Ruhr-Universität Bochum, Bochum, September 1983.

[8] Guibas, L. J., Kung, H. T., Thomson, C. D.: Direct VLSI Implementation of Combinatorial Algorithms. Research Report, Carnegie Mellon University, Pittsburgh, 1979.

[9] Katona, E.: Cellular Algorithms: Selected Results of the Research Group on Automata Theory Szeged, von Neumann Society, Budapest, 1981 /in Hungarian/ 165 pages.

[10] Katona, E.: Cellular algorithms for Binary Matrix Operations. Proceedings of the conference CONPAR'81. Nürnberg, 1981, Lecture Notes in Computer Science 111, Springer, 1981, pp. 203-216.

[11] Katona, E.: Local and Global Reversibility of Finite Inhomogeneous Cellular Automaton. Acta Cybernetica, 3, 1978, pp. 287-292.

[12] Katona, E., Legendi, T.: Cellular Algorithms for Fixed Point Decimal Addition and Multiplication. Elektronische Informationsverarbeitung und Kybernetik, EIK 17, 1981, pp. 637-644.

[13] Katona, E.: A Design Methodology for Complex Cellular Programs. Presented at II. Int. Workshop on Parallel Processing by Cellular Automata and Arrays, Berlin, 1984.

[14] Kovács, I.: Cellular Program for the Sobel Transformation on a Traditional Component Based Cellprocessor. Presented at II. Int. Workshop on Parallel Processing by Cellular Automata and Arrays, Berlin, 1984.

[15] Köles, P.: From Cellular Algorithms to Cellular Programs. Presented at II. Int. Workshop on Parallel Processing by Cellular Automata and Arrays, Berlin, 1984.

[16] Kung, H. T.: Let's Design Algorithms for VLSI Systems. Report No. CMU-CS-79-151, Carnegie-Mellon University, Pittsburgh, 1979.

[17] Kung, H. T.: Why Systolic Architectures?, Computer, Jan, 1982, pp. 37-46.

[18] Kung, H. T.: The Structures of Parallel Algorithms. Advances in Computers, 19, 1980, pp. 65-112.

[19] Kung, H. T.,-Leiserson, C. E.: Algorithms of VLSI Processor Arrays. Symposium on Sparse Matrix Computations 1978, Proceedings, Knoxville, 1978.

[20] Kung, H. T.,-Ruane, L. M., Yen, D. W. L.: A Two-level Pipelined Systolic Array for Convolutions. VLSI Systems and Computations, Computer Science Press, 1981, pp. 255-264.

[21] Legendi, T.: Programming of Cellular Processors. Informatik-Berichte Nr. 7703, T. U. Braunschweig, 1977, pp. 53-66.

[22] Legendi, T.: INTERCELLAS- an Interactive Cellular Space Simulation Language. Acta Cybernetica, 3, 1977, pp. 261-267.

[23] Legendi, T.: A 2D Transition Function Definition Language for a Subsystem of the CELLAS Cellular Processor Simulation Language. Computational Linguistics and Computer Languages, 13, 1979, pp. 169-194.

[24] Legendi, T.: Cellprocessors in Computer Architecture. Computational Linguistics and Computer Languages, 11. 1977, pp. 147-167.

[25] Legendi, T.: Cellular Algorithms and their verification. Invited paper. Proceedings of the conference CONPAR'81, Lecture Notes in Computer Science 111, 1981, pp. 169-188.

[26] Legendi, T.: Implementation and Application of Cellular Processors. Proceedings of the conference Parallel Computing '83, Elsevier Science Publisher B. V. /North-Holland/, 1984.

[27] Parkinson, D.: The ICL Distributed Array Processor. Infotech Future Systems Vol. 2, Infotech International, Maidenhead Berkshire, 1977, pp. 389-402.

[28] Snyder, L.: The Blue CHiP Project. A Summary of Current Activities. Purdue University, 18 pages.

[29] Snyder, L.: Introduction to the Configurable Highly Parallel Computer. Computer, 15, 1, 1982, pp. 47-56.

[30] Tóth, J.: Design of a Cellprocessor Model. Presented at I. Int. Workshop on Parallel Processing by Cellular Automata, Berlin, 1982.

[31] Tóth, J.: A 64 Kcell LSI Based Cellprocessor Prototype /SPP/, Presented at II. Int. Workshop on Parallel Processing by Cellular Automata and Arrays, Berlin, 1984.

[32] Vollmar, R., Legendi, T. /eds/: Cellprocessors and Cellgorithms. Report on a Common Research Project, Informatik Skripten No. 1., T.U. Braunschweig, 1981.

[33] Zsótér, A.: Cellprocessor Based on Traditional Components. Presented at II. Int. Workshop on Parallel Processing by Cellular Automata and Arrays, Berlin, 1984.

THE SOFTWARE ENVIRONMENT OF THE CELLULAR PROCESSOR MODEL

Barsi, I.[1/]

Preface

The main goal of the research on cellular automata at Szeged is to build up economically a cellprocessor under the given technological conditions. Considering primarily the economical viability, Legendi, T. proposed [1] the cellprocessor architecture using microcells in 1976. This proposal and other works e. g. [2], also contain the basic components of the software environment that drive and supervise the operation of the cellprocessor to be suggested. Though the foundations had been established, before the realization of a cellprocessor which is capable of computing more considerable tasks, it was expedient to develop and build a relatively small cellprocessor model with the help of which partly the operation of a cellprocessor architecture partly the behaviour of the already realized cellprograms could be studied in practice.

Taking the above into account, the cellprocessor model whose implementation is in its final phase was developed by Tóth, J. in 1983 [3].

The purpose of this paper is to present the software environment of this cellprocessor model. However, before we would turn our attention to this question, we present the cellprocessor itself in a few words. The detailed discussion can be found in the related works listed in the References at the end of this paper.

The Cellprocessor Model

The model is a specialized peripheral device which augments the efficiency of the host machine to which it is connected in performing tasks that can be parallel-processed well. The structure of the model and its connection to the host machine are shown in figure 1. The components of the cellprocessor and their functions are as follow:

HOST machine: It holds the user programs and the handler through which the user program can call the cellprocessor so that it can perform the computation required. The host machine is a micro-computer based on a Z-80 microprocessor. A draft specification of it is:

1/Department of Computer Science, JATE University, Szeged
Somogyi u. 7., H-6720 Szeged, Hungary

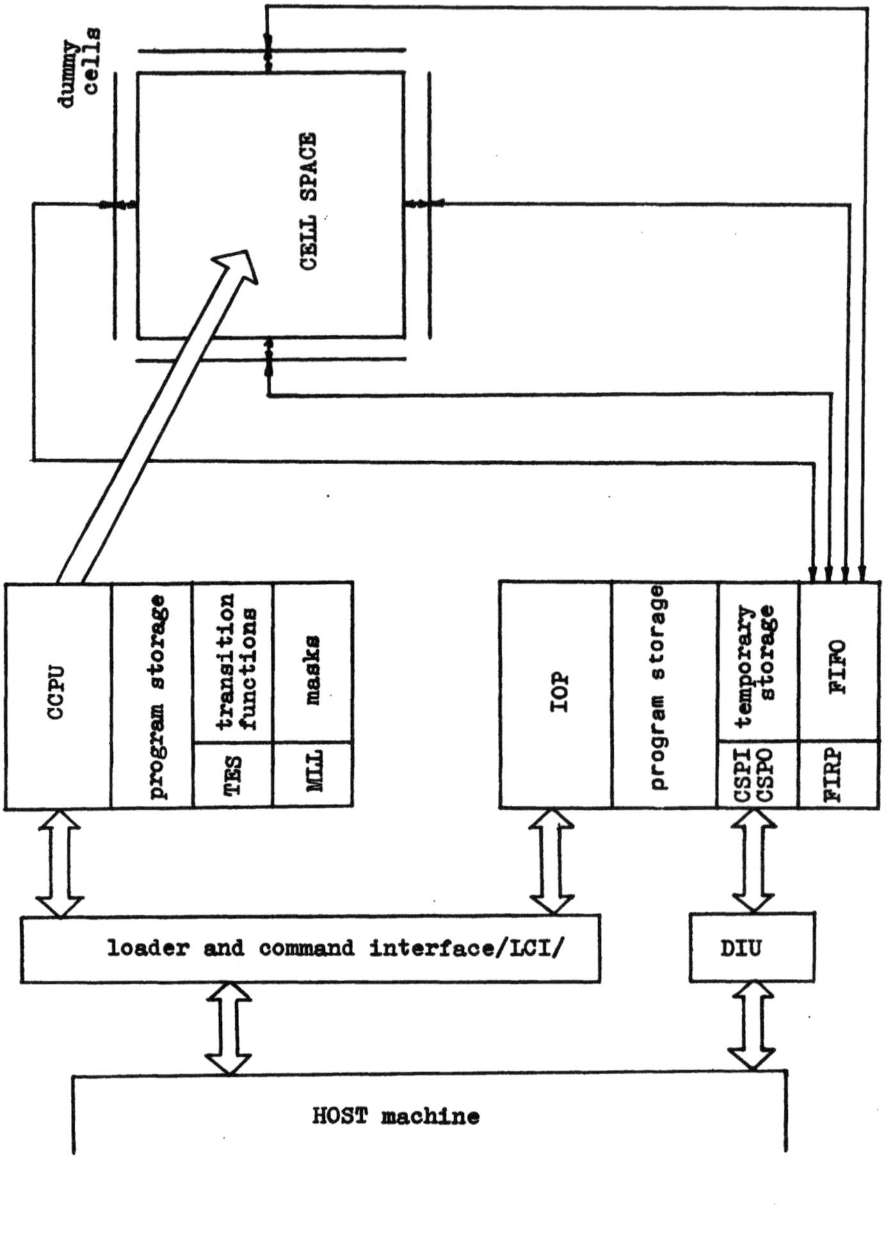

Figure 1. The cellprocessor model and the host machine

Operation system: Interactive Disk Operation System.
Main software components: Z-80 assembler, high level program languages /BASIC, PASCAL, PL-MICRO/, MP/3 macroprocessor, EDITOR program, utility programs.
Technical parameters: 64 K RAM, floppy disk drive, movable disk drive, line printer, data transmission line.

LCI - Loader and Command Interface. Cellprograms and control commands are loaded into the cellprocessor through this interface.

CCPU - Central Cell Controller Unit. One of the most important parts of the cellprocessor model is the CCPU. The CCPU program component is loaded into the program storage. This program controls the execution of the transition functions with the help of the TES /Term Sequencer/ and the MLL /Mask Loading Logic/ subprocessors. The masks designate the cells for which the given transition function is valid, realizing in this manner the inhomogeneous cell space.

DIU - Data Interface Unit. The DIU establishes a direct connection between the host and the IOP using the DMA /Direct Memory Access/ facility.

IOP - Input Output Processor. The otner part of outstanding importance within the cellprocessor is the IOP. The program storage of the IOP holds the IOP program component of cellprograms. The task of this program is to control the input/output processes between the host machine and the cellprocessor. The CSPI /Converting Subprocessor Input/ and the CSPO convert the input/output data into the right format needed by the cellprograms /cellgorithms/ and the user program. From the CSPI temporary storage, the already converted data go to the FIFO-s, and from here to the cellspace. The result of computation enter one of the FIFO-s again and from here the data go to the CSPO temporary storage after the right format-conversion.

A cellprogram is made up of four components:

- masks,
- transition functions,
- CCPU program component, and
- IOP program component.

The cellprogram is catalogued in a library on a disk device connected to the host machine. When invoking the cellprogram, the software which resides on the host machine loads the parts of the cellprogram into the appropriate units of the cellprocessor model.

The Software Environment

In a wider sense, the supporting host machine's entire software system belongs to the environment because without this any kind of applications is unthinkable. Within the frame of this paper, however, we are interested only in the software system that has a close connection with the cellprocessor model. This software environment is made up of five components:

- CEDITOR editor program,
- SPMASS assembler program,
- SPMSIM simulator program,
- SPM MACRO LIBRARY, and
- BASIC extension.

CEDITOR Program

This program facilitates the writing and preparing of the cellprograms. Its interactive mode of operation and many advantageous features make the programmer's work easier.
The program is suitable for writing all the components of a cellprogram.

- Masks.
  In consequence of the small cellular space size, the specification of masks can be carried out in a natural way, which means that, according to the actual cellspace arrangement, a 8*32 or a 16*16 matrix appears on the screen for each mask and so the programmer can easily give the mask required. The result of the editing is a so called M type file.
- Transition Functions.
  Each function is composed of terms. Each term is coded in a byte and consits of the following:

  Ø J K N E S W

  where:

  J,K - a flip-flop value of the given cell, which defines the state of the cell.

  N,E,S,W - designate the state of the north, east, south and west neighbour of the cell respectively.

With the help of the program the operator can enter these terms for every transition function. The editing results in a so called F type file.

- CCPU and IOP program components.
  As to their statements and directives the format of the program lines are conform to the generally used other assembly language formats. The CEDITOR program knows the mnemonics of the statements and directives, and helps the work of the programmer by not allowing to enter syntactically erroneous lines. The result is the so called C resp. I type file.

SPMASS Assembler Program

From the M, F, C, I type files, the assembler generates a so called S type file that consists of all the components of a cellprogram in loadable form. Figure 2. shows the process of development of a cellprogram. The loadable form also contains additional information that are necessary for loading the components into the model. This format is shown in Figure 3. The S type file is stored and catalogued by the assembler in a library. The execution of the cellprogram can be carried out either on the cellprocessor model or by the SPMSIM simulator program.

SPMSIM Simulator Program

The simulator program has a very great importance both in debugging the cellprograms and at the examination of the expectable behaviour of the cellprocessor model. And this especially holds true in the latter case because the hardware is of a very new type as to its structure and functions.

The simulator which closely follows the structure and the operation of the model in its structure and operation affords a great help from cellprogramming and hardware development standpoints. The program can operate in two basic modes of operation:

- HAND mode and
- SUB mode.

In HAND mode, the execution of cellprograms is performed under the interactive control of the user. The operator has to designate the name of the cellprogram to be run, the addresses of the I/O areas, the control commands, etc. There is a possibility to insert break points and to run the simulator by basic clock periods. In both cases, all parts of the simulated cellprocessor model /registers, status words, cell-

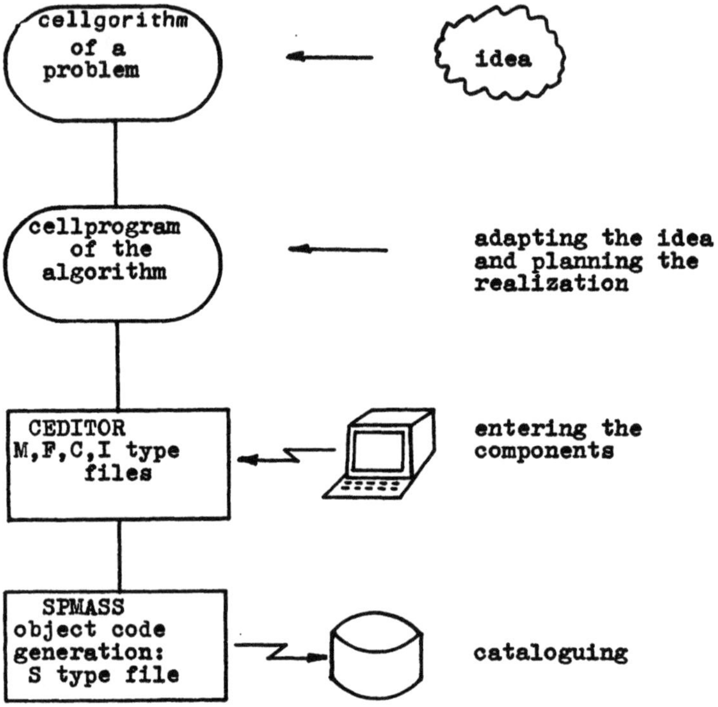

Figure 2. The process of cellprogram development

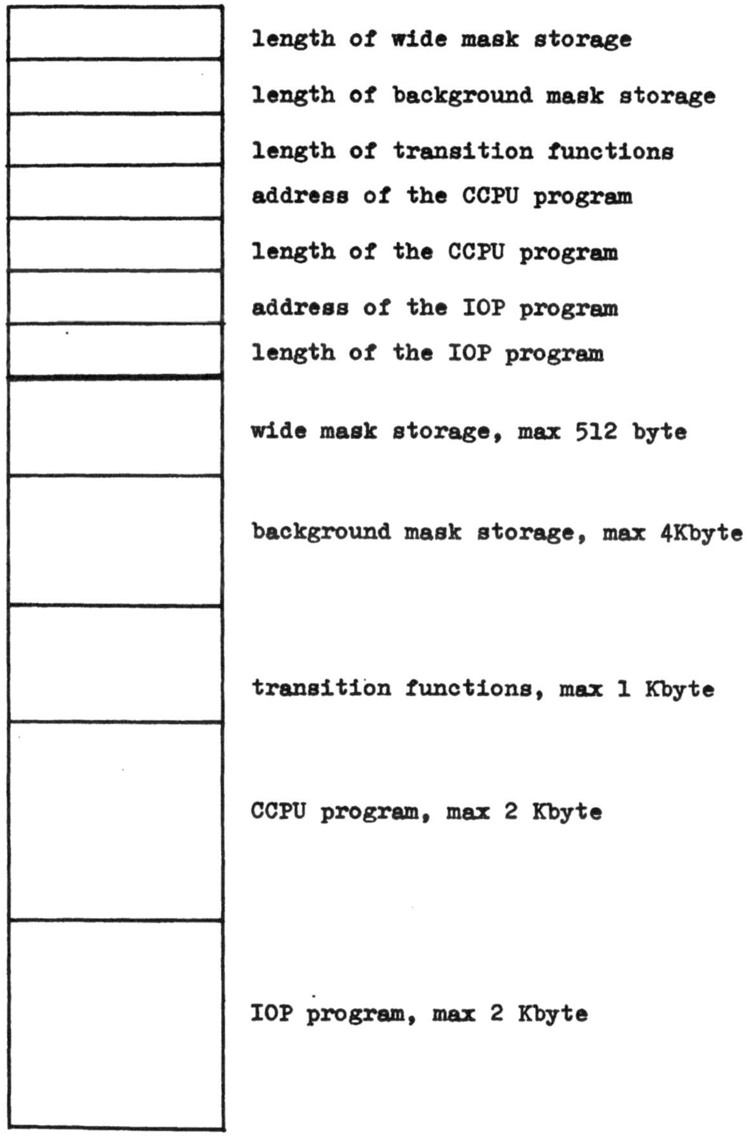

Figure 3. The structure of the S type file

space, etc./ can be displayed and if necessary, the important information can be printed out.
In SUB mode, the cellprograms can be executed from user programs. In this case, the parametering takes place via a control block which consists of the description of the I/O areas, the name of the cellprogram to be run and in coded form the control commands that specify the state of the simulated cellprocessor for further running after a stop condition has occured.

SPMHNDLR Handler

The task of the handler is to control and supervise the operation of the cellprocessor model. The reception of parameters which are needed to run a cellprogram is performed through a control block. With the help of the control block, the handler performs the following tasks:

- searching the disk for the specified S type file /cellprogram/,
- according to the concrete call, putting the actual parameters into the possibly occured postdefinite variables,
- starting the cellprogram and controlling its run on the basis of the control list,
- managing the IT-s generated by the cellprocessor mode,
- error management, that is attempting to correct the errors if possible, or suspending the further run of the cellprogram in case of severe errors and sending error messages to the caller.

Figure 4, shows the relationship between the handler and the user program and the cellprocessor model.
The handler has a so called master mode of operation too. This mode of operation enables the user to call and parameter the handler directly from a user written program. This option supports mainly testing purposes. Using this option the user is capable of writing and reading the various parts of the cellprocessor or running it even for a basic clock period.

SPM MACRO LIBRARY

In order to run a cellprogram, the user has to build a control block in his assembly language program, which provides the necessary informations for the handler. The generation of such a control block requires a great attention and the full knowledge of the operation of the cellprogram invoked. To reduce the possible error sources and so that the user should not take care of control block building, a macro library

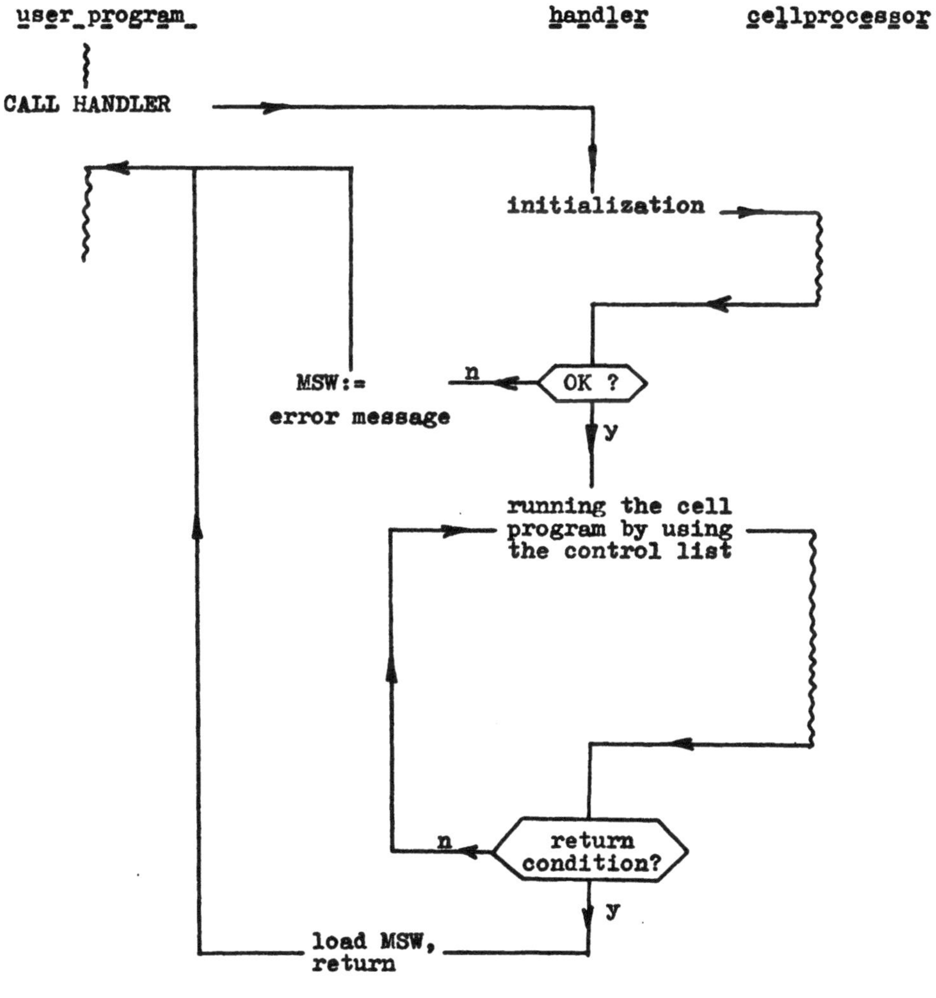

Figure 4. The relationship between the handler, the user program and the cellprocessor model

is at the user's disposal. After the suitable parametering of the required macro, the statements which perform the handler call and the control block which holds the required informations are generated during the macro expansion. This is shown in Figure 5.

BASIC Extension

We want to make it possible that the advantages offered by the cellprocessor model should be taken not only from an assembly language program but also from the level of higher program languages. The possibilities on how to run cellprograms within the frame of the software environment of the cellprocessor model is given in Figure 6. First we realized the Z-80 BASIC language extension. The extension is conform to the formats generally applied to the statements of BASIC programs. The user can decide of his own free will whether he takes or not the advantages of the cellprocessor model at the solution of a problem. The format of the call is:

serial number CLET variable=expression

Here the operation codes designate the call and the execution of cellprograms. Because of the small size of the model, defining the operations our aim is not the completness but the demonstration of the fact that the cellprocessor, in general, is a powerful tool for solving certain problems.

Up to now the following operations have already been realized or are being under realization:

- product of integers,
- product of a vector and an integer,
- product of two bit matrices,
- sum of integers,
- sum of two general matrices,
- sorting,
- associative memory,
- image processing operations, as
    - . enlargement,
    - . reflection,
    - . rotation.

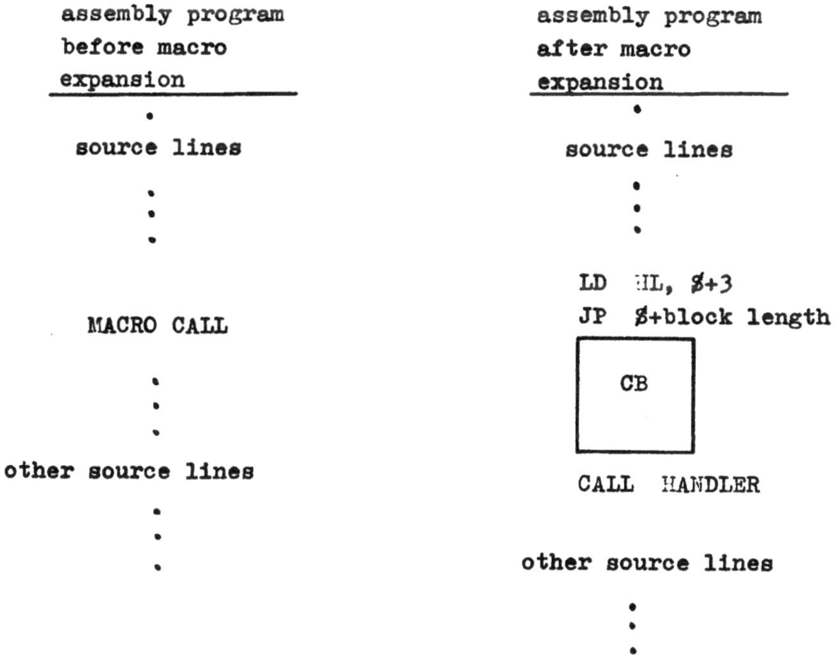

Figure 5. Using the cellular macros

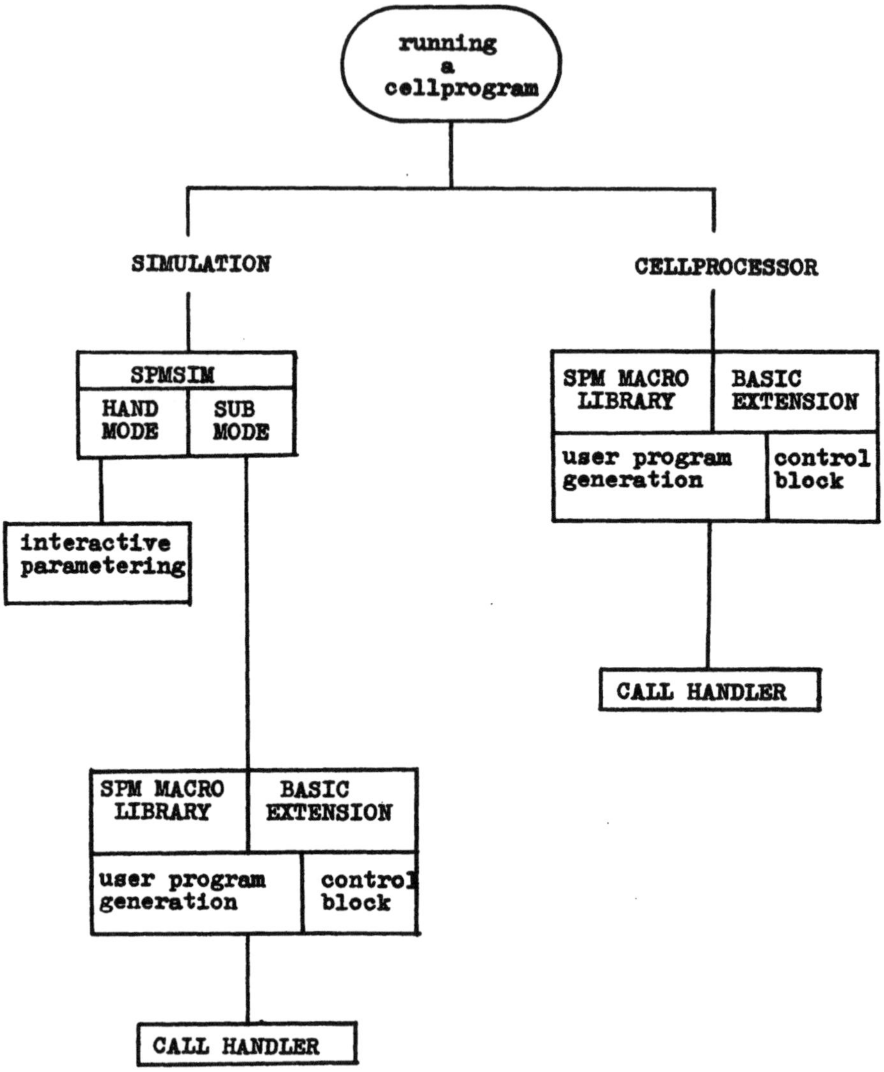

Figure 6. How to run cellprograms. Summary

Summary

The main goal and the concept of the Cellprocessor Model software environment project have been discussed, the design and the implementation of the system shown.

Some components of the system, the CEDITOR, SPMASS, SPMSIM, SPM MACRO LIBRARY and BASIC extension, and their main features have been presented.

The discussion presents the cellprogram development process, shows that the system offers a convenient use of the model, and gives a prototype for software systems of cellprocessors.

References

[1] Legendi, T.: Planning, Programming and Applying Parallel Operating Homogeneous Systems. Conference on "Programming Systems'78" Szeged 1976 /in Hungarian/.

[2] Legendi, T.: Cellprocessors and Cellgorithms - Report on a Common Research Project - edited by R. Vollmar and T. Legendi. Braunschweig 1981.

[3] Toth, J.: Design of a Cellprocessor Model, PhD dissertation in Hungarian. TU of Budapest, EE Faculty, 1983.

[4] Toth, J.: Documentation of the SP Model, I., II., III. volumes, in Hungarian. Research Group on Automata Theory of the Hungarian Academy of Sciences. Szeged 1983.

[5] Toth, J.: The SP/M Microassembly Language /manuscript in Hungarian/. Research Group on Automata Theory of the Hungarian Academy of Sciences. Szeged 1983.

PARALLEL FINITE ELEMENT ALGORITHM AND THE RESPECTIVE CELLULAR STRUCTURES

L. Bernáth[1/]

S u m m a r y

Finite Element Method is a highly efficient means for solving a wide range of practical engineering problems such as mechanical deformation, strength distribution in solid bodies of complex shape, heat-conduction problems, electromagnetic-field analysis, fluid flow, problems of non-stationary continuums. In certain cases there is simply no other way at all to predict the behavior of an object under irregular load conditions. However, this excellent method is rather expensive because of its need for large computing capacity.

In this paper we demonstrate the application possibilities of special cellular architectures as a proper means for computing FE problems. First, we point out that the FE-principle is locally parallel natured. The common joints /nodes/ of the neighbouring elements can be considered as cells, the actual state of which are exclusively determined by the latest states of the respective neighbouring cells. We also give the transition function for a certain cell, according to which its new state will be determined in the next cycle. This way we defined a parallel iterative algorithm for FEM.

The next part of the paper demonstrates the principle of a proper cellular architecture to realise the algorithm above.

We come to the conclusion, that FEM might be an application field of major importance for cellular computing.

1. Introduction

As a result of the wide-scale, long theoretic preparation work at the Research Group on Automata Theory headed by Dr. Legendi, now we have got into the position to develope and build pilot-models of specific cellular processors suitable for solving certain computing problems of parallel manner. Obviously, a fundamentally new and unusual device like a cellular processor can not comply with the requirements unless the particular demands and peculiarities of the perspective application fields are taken into account even at the earliest stage of development. As a question of significance it is important to point out: how we can fully utilize the advantages of a cellular architecture

[1/]Department of Computer Science, JATE University Szeged
Somogyi u. 7., H-6720 Szeged, Hungary

by formulating the problem to be solved from parallel aspect. This parallel aspect might seem us rather unusual, however most of the physical phenomena and real processes are fundamentally parallel natured. Optimally, the structure of the algorithm, the mathematical model of a certain problem, and that of an appropriate cellular processor cover each other well. This idea will be demonstrated in a short account on our efforts to elaborate a finite-element algorithm and the respective cellular structure.

2. The essence of FEM

The so called finite-element method is a highly efficient means for solving practical engineering problems such as mechanical stress- and deflection analysis of complex shapes subject to heating and external loads, stiffness and stability problems in structural and continuum mechanics, fluid flow, heat conduction, thermal deformation, non-stationary temperature distribution and vibration analysis etc. Complex as the structure or the boundary condition is, the accuracy of the method can be as high as needed. No wonder that this general, numerical procedure is considered one of the major part of an up-to-date CAD system. In certain cases there is simply no alternative method at all to predict the behavior of a nonlinear, complex shaped body under non-stationary external or internal load. For example just think of a body of a car clashing against a massive wall. An appropriate FE model and the resulting computer simulation will help the constructor to study deformations step by step without expensive, time-consuming real-life experiments. The only disadvantage of the method is its need for a hudge computing capacity. Process times of several dozen houres are not extreme. Even this circumstance justifies our effort to develop a cellular FE algorithm and a suitable processor. The FE- principle is fundamentally parallel, as you will see, so we have hopes of finding and algorithm suitable for cellularprocessing.

The point of FE-idea is as follows: /see FIG.1/
A geometrically complex shaped, inhomogenous, continuous body, for example an engine part, can be replaced with an idealised mesh of geometrically simple shaped, homogenous elements. The neighbouring elements in the mesh are joined with each other only at their common nodes. Displacement field inside a certain element, as a result of its regular geometry and homogenity, can be described as a relatively simple, mostly linear function of the nodal-displacements of the respective element and the external or internal load having a direct effect on the same element. Consequently, knowing the nodal-displacements, the

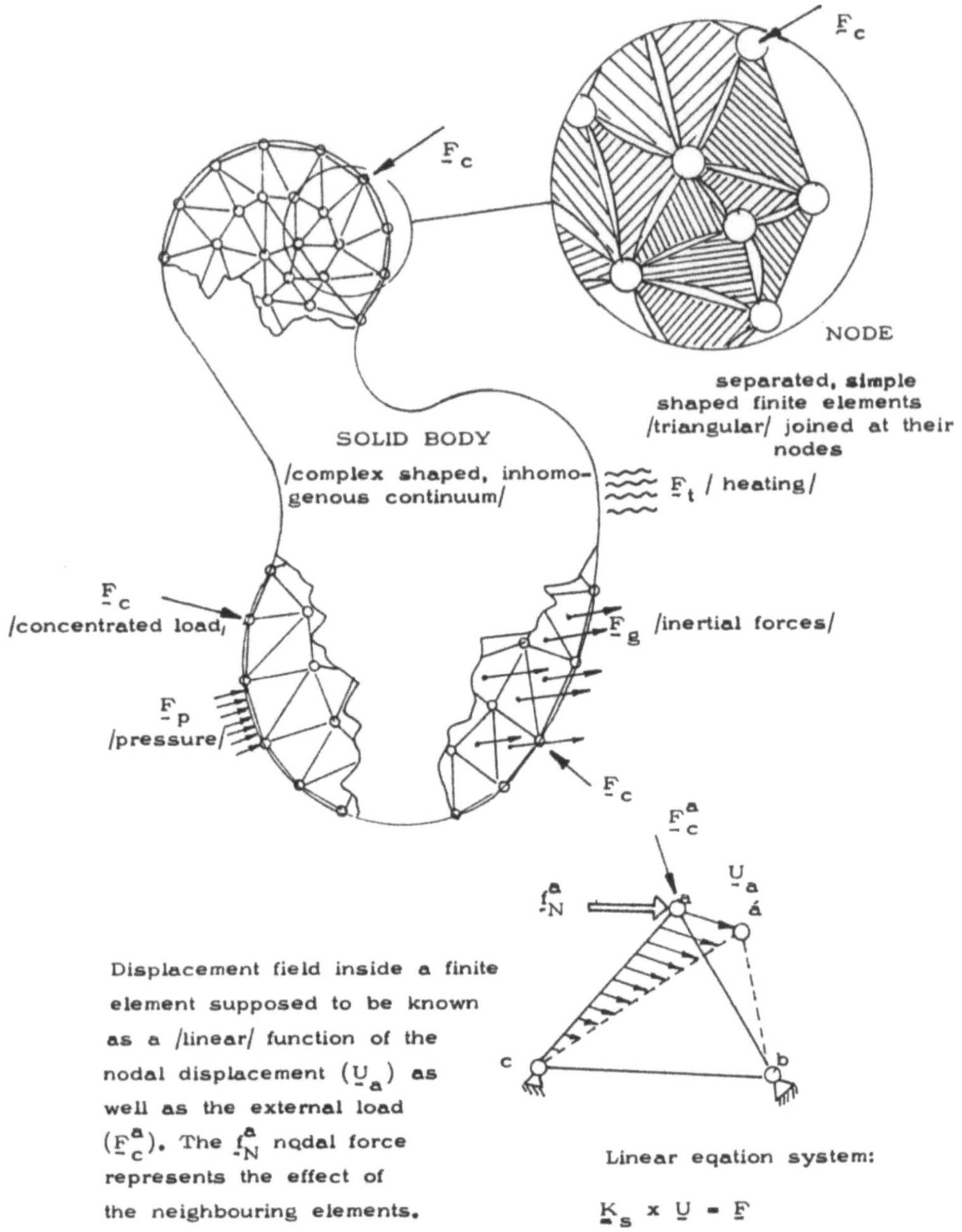

Fig. 1 Finite element idealization of a complex shaped solid continuum subject to external and internal loads ($\underline{F}$)

displacement field of the whole structure can be determined, what is our main aim. So that nodal-displacements are considered primary unknown data in the FEM. In order to determine nodal-displacements as a result of the external and internal load effecting on the structure, the so called "elementary stiffness matrix" ($\underline{\underline{K}}_e$) is to be calculated. This matrix describes the linear connection between nodal-displacement and external load, concerning a single element. The value of its components depends on the geometry and characteristics of the respective element. These factors are known.
The next step to follow is to formulate the so called "connection matrix" ($\underline{\underline{C}}$). The connection matrix defines the relations among the elements, describes the structure of the mesh in matrix form. As a result of the very first phase of FEM, called mesh generation we can get the connection matrix just simply. The next step to follow is a composition of the elementary stiffness m. and the connection matrix to get the so called "system stiffness matrix" ($\underline{\underline{K}}_s$). This matrix represents the linear function between the external /and internal/ load vector ($\underline{F}$) and the unknown nodal-displacement vector ($\underline{U}$) in form of a linear equation system. The number of unknown data in this equation system may be as many as ten thousand, so its solution takes a lot of time using sequential computer.

3. Cellular FE algorithm

How could we transform this sequential FE algorithm to be suitable for parallel computation? /see FIG. 2/ Let us investigate a certain node in the mesh more thoroughly! It is easy to notice that the unknown displacement of a certain node is a linear function of the displacements of the neighbouring nodes. We define two nodes in the structure as "neighbours" if there exist an element containing both of them. Nothing else but the displacement of the neighbouring nodes effect the displacement of a certain node. Nodal-displacement caused by the external load, effecting the node itself directly, is defined as "initial displacement".
Now, let us regard the nodes as cells, and the nodal-displacements as states of the respective cells! A linear combination of the former states of the neighbouring cells, superposed by the initial state will describe the transition function for each cell. For the whole cellular field, the transition function can be given in matrix form. Obviously, this is nothing else but an iterative formulation of the FE idea from cellular aspect. The iterative transition function and the conventional linear equation system are equivalent. The matrix of the transition

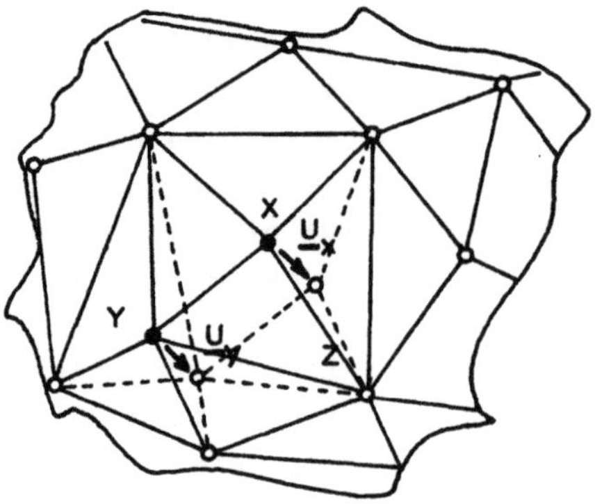

Nodes X and Y are "neighbours" since element e_{xyz} involves both of them.

Linear connection between nodal displacements $\underline{U}_x$ and $\underline{U}_y$

(1) $\underline{U}_x = \underline{b} * \underline{U}_y$, where

$\underline{U}_x$ - the unknown displacement of node X, caused by

$\underline{U}_y$ - the given displacement of node Y

$\underline{b}$ - constant stiffness vector

- - - - - - - -

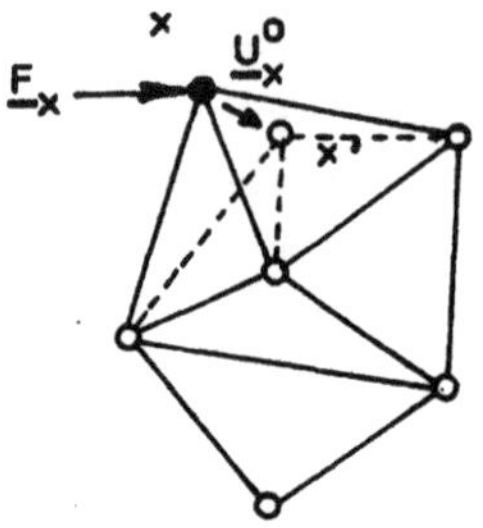

Initial displacement of a certain node, caused by the external force $\underline{F}_x$ having direct effect on the very same node.

(2) $\underline{U}_x^o = \underline{b}_o \times \underline{F}_x$

The general formula for a certain node X can be determined by a superposition of (1) and (2):

$$\underline{U}_x = \underline{b} \times \underline{U}_y + \underline{U}_x^o$$

Let us take all neighbours into account: $\underline{U}_x = \underline{\underline{B}} \times \underline{U}_{\Sigma y} + \underline{U}_x^o$, where

$\underline{\underline{B}}$ - the stiffness matrix for X

$\underline{U}_{\Sigma y}$ - the displacement vect. of all neighbouring nodes of X

For the whole system of finite element structure the formula can be given in ITERATIVE form:

/n- is the cycle index/

$$\boxed{\underline{U}^{n+1} = \underline{\underline{B}} \times \underline{U}^n + \underline{U}^o} \qquad (3)$$

FIG. 2

Cellular representation of the finite element idea

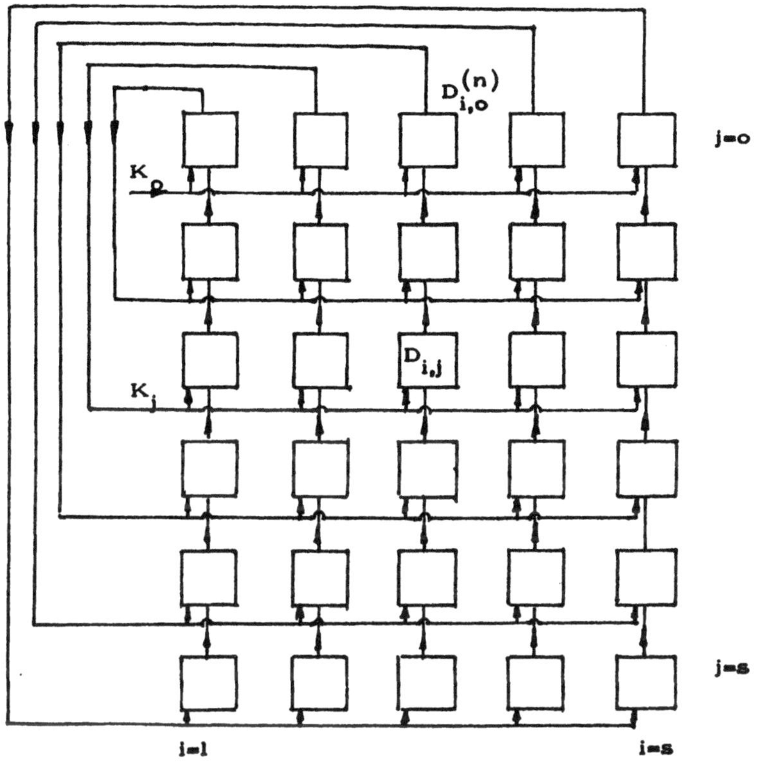

s- the number of unknowns

n- cycle index

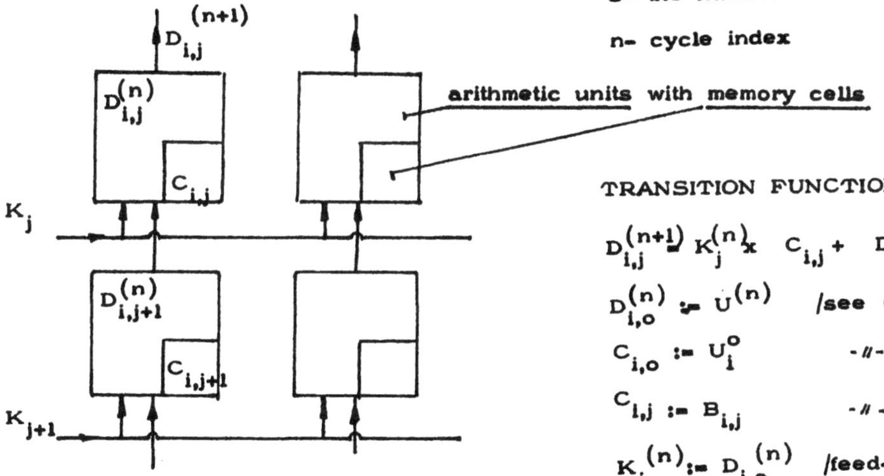

TRANSITION FUNCTION

$D_{i,j}^{(n+1)} := K_j^{(n)} \times C_{i,j} + D_{i,j+1}^{(n)}$

$D_{i,o}^{(n)} := U^{(n)}$ /see (3) /

$C_{i,o} := U_i^o$ -//-

$C_{i,j} := B_{i,j}$ -//-

$K_j^{(n)} := D_{i,o}^{(n)}$ /feed-back/

$K_o = 1$

FIG. 3
Cellular architecture for parallel FE algorithm (3)

function can be derived from the "system stiffness matrix".

4. A cellular architecture

A special cellular architecture has been worked out to realise the algorithm above. The schematic figure demonstrates a vertically structured cellular field of arithmetic units /"cells"/. The number of columns equals the number of unknowns - i. e. nodal displacements - in the iteration formula. Each unit calculates its own actual state $\left(D_{i,j}^{n+1}\right)$ according to the same transition function. The main characteristic of the structure is the feed-back type, continous data-flow allowing a high-speed parallel computation. Each column, as a chain of arithmetic cells, generates a new walue of the respective iterative unknown, represented by the actual state of the cell in the first line of the column. The whole procedure is monitored by a controll - system to block the iteration provided that the so called "stop-criteria" comes true. In that case the actual states of arithmetic cells in the first line give the iterative solution to the original FE problem. /see FIG. 3/

5. Conclusion

Finally, we came to the conclusion that FEM might turn up to be one of the major application fields of parallel computing and cellular processors. Though at the present state of technology the cellular architecture demonstrated above seems rather unrealistic to build up in fact, the rapid development of microelectronics and particulary the latest events in wafer-scale integration might allow us to realise similar processors in the near future.

Bibliography

/1/ Zienkiewicz, O.C.: The FEM in Structural and Continuum Mechanics. /London, 1977/

/2/ Hinton E.- Owen D.: An Introduction to FE Computation. /London, 1981/

/3/ Stabrowski M.M.: An Algorithm for the Solution of Very Large Banded Unsymmetric Linear Equation Systems. /IJNME vol. 17 p: 1103-1117, 1981/

/4/ Wennerström, H.: Numerical and Computer Techniques in FE Analysis. /Göteborg, 1981/

/5/ Rao, S. S.: The FEM in Engineering /Oxford, 1982/

/6/ Nguyen van Phai: Automatische Netzgenerierung für Dreidimensionale Festigkeitberechnungen mit der FEM. /Kaiserlauten 1979/

A DESIGN METHODOLOGY FOR COMPLEX CELLULAR PROGRAMS

Endre Katona [1)]

Abstract. Complex cellular algorithms can be described by a graph the nodes of which represent basic operations (addition, multiplication, comparison, etc.) and the edges of which mean data transmissions. The graph describes synchronous computation (it is not a data-flow graph). Such a graph can be embedded into the cellular field of a universal-purpose cellprocessor by replacing the nodes by cellular processing elements and the edges by data channels. This paper gives an exact definition of the cellular program graph, illustrates its practical use with an example and shows the possibility of automatic translation from a high-level language into a cellular program.

1. CELLULAR PROGRAMMING

A cellular processor is based on a homogeneous net of elementary processors. The programming of such an architecture drastically differs from that of a sequential or multiprocessor computer. Array processors are in some sense similar to cellular processors. However, they are usually built from arithmetic processors (macro-cells) operating on word-level, while - in our terminology - a cellular processor consists of bit-processors (micro-cells) performing bit-level operations. As a consequence, basic arithmetic operations should be realized in cellular software if a cellprocessor is applied. This programming difficulty is compensated by two advantages:

- the basic operations can be executed in a bit-parallel way;
- there is no fixed word length in the cellprocessor (long words can be processed with the same efficiency as short ones).

In this paper a <u>universal-purpose cellprocessor architecture</u> is considered [2]. Here we give a rough sketch of it only, for details see [4].

Our cellprocessor architecture consists of 3 components (fig.1). The <u>cellular field</u> is a two-dimensional array of elementary processors (cells), the <u>CCU</u> (cellprocessor control unit) sends a microprogram to the cellular field, and the <u>I/O processor</u> ensures data-transmission between the host computer and the border of the cellular field.

---

1) Research Group on Automata Theory
Hungarian Academy of Sciences
Somogyi u. 7, H-6720 Szeged, Hungary

The cells in the cellular field are connected locally, but the architecture offers a special switching network ensuring farther connections [4]. This latter ability of the cellprocessor - called teletransport - is an important help in cellular programming.

Each cell in the cellular space has two registers, i.e. an external state register for processing and an internal state register ensuring inhomogeneity. Both registers contain 1-6 bits only, depending on the concrete architecture.

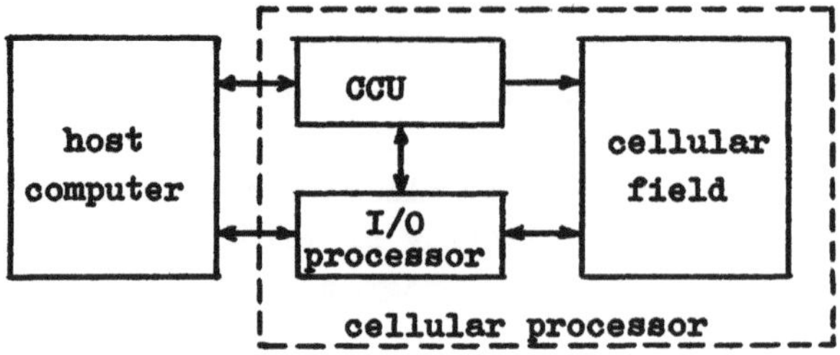

Fig.1. Sketch of a cellular processor.

In the course of cellular programming processing areas are allocated in the cellular space, called cellular processing elements (cellular PEs), which are interconnected by shifting areas, called cellular data channels (fig.2). Both the PEs and data channels are realized in software by loading a suitable pattern of internal states into the cellular field (fig.3). The cellular field is totally reconfigurable. The notions "cellular PE" and "cellular data channel" require some detailed explanation as given below.

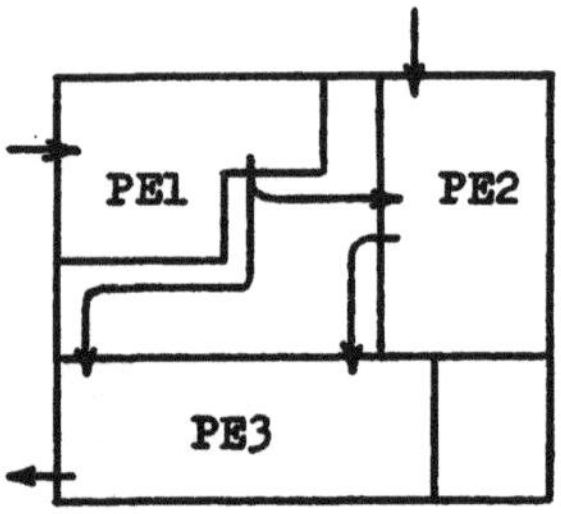

Fig.2. The structure of a simple cellular program consisting of 3 PEs (an ad hoc example).

1	1	1	1	1	0	2	2	2
1	1	1	1	1	0	2	2	2
1	1	1	4	6	6	2	2	2
1	1	1	4	0	5	2	2	2
5	5	5	4	0	4	2	2	2
4	0	0	0	0	4	2	2	2
3	3	3	3	3	3	3	0	0
3	3	3	3	3	3	3	0	0
3	3	3	3	3	3	3	0	0

Fig.3. The internal state configuration of the cellular program given in fig.2. The cells with internal states 4, 5 and 6 shift the external states down, left and right, respectively.

A cellular PE is a group of cells performing certain elementary computation. The PE gets an input data stream from the neighboring cells and sends an output data stream to the neighboring cells (fig.4), where the data $y_1^{(i)},\dots,y_m^{(i)}$ are computed from $x_1^{(i)},\dots,x_n^{(i)}$ for any i.

For example, if an adder is considered, then $y^{(i)} = x_1^{(i)} + x_2^{(i)}$ (n=2, m=1).

We say that the PE of fig.4 has n input operands, $x_1,\dots,x_n$, and m output operands, $y_1,\dots,y_m$. The <u>succeeding time</u> of an operand x_j is defined as the number of steps between the entry of the first bit of $x_j^{(i)}$ and $x_j^{(i+1)}$. The succeeding time is supposed to be the same for all input and output operands of a PE. If two PEs are interconnected by a data channel, they should have the same succeeding time, too.

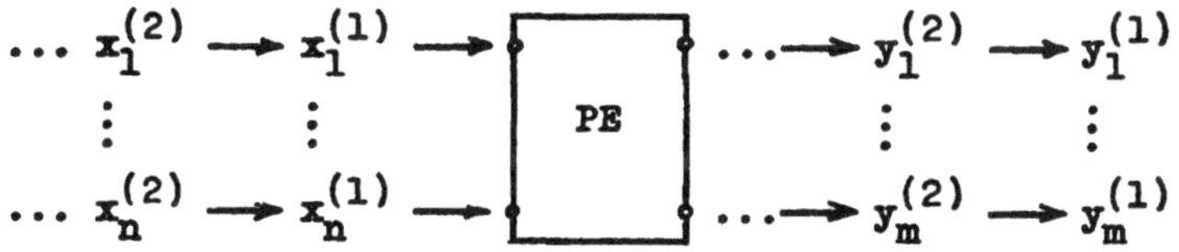

Fig.4. A cellular PE processes a continuous data stream.

A <u>cellular data channel</u> is a sequence of "neighboring" cells, where "neighboring" means either real neighbors or cells connected by teletransport. The external states are shifted from cell to cell inside the data channel.

2. CELLULAR PROGRAM GRAPHS

A cellular program (fig.2) can be described by a graph the nodes of which represent PEs and the edges of which represent interconnections (data channels). An exact definition of such a graph is given below [1)]. Note that a similar model has been used in [1] for VLSI circuit design; the similarity between circuits and cellular programs will turn out in many points in the further discussion.

A <u>cellular program graph</u> (CP-graph) is a special directed graph with three types of nodes:

- PE-nodes,
- storage nodes,
- IOP node.

In hardware thinking storage nodes correspond to registers, PE-nodes to combinational logic, and the IOP node represents the external connections of the system [1].

A cellular program interpretation of the graph is as follows.

<u>PE-nodes</u> represent cellular processing elements. If a cellular PE has n input operands, then the corresponding PE-node has n <u>input points</u> and m <u>output points</u>. PE-nodes are denoted by squares, the input and

---

1) A formal mathematical definition of the CP-graph would be rather complicated, therefore we have decided to give an informal but exact description.

output points are marked by small circles on the edges of the square.

Storage nodes represent delays required in the cellular program. They are implemented either explicitly by data channels or implicitly by the delays of cellular PEs. Storage nodes are denoted by circles.

The IOP node represents the I/O processor of the cellprocessor. In our notation the IOP node does not appear in the CP-graph, but all edges from or to the IOP node are denoted by double arrows while other edges by single arrows. This notation is explained in fig.5.

Fig.5. The CP-graph of the cellular program given in fig.2.

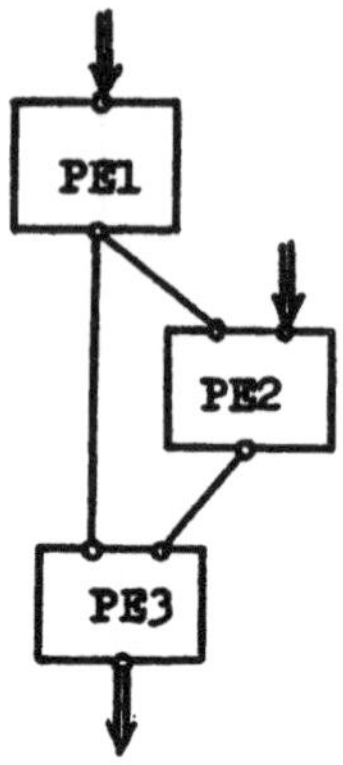

Rules of constructing a CP-graph:

- Any input point of a PE-node must be the head of exactly one edge.
- Any output point of a PE-node may be the tail of arbitrary many edges.
- A storage node may be the head of maximum one edge and the tail of arbitrary many edges.
- Any cycle in the CP-graph must contain a storage node. Without this assumption an infinite ripple could arise in the CP-graph.

CP-graphs define parallel computation as well as data-flow graphs. However, a CP-graph performs synchronous computation, similarly to a synchronous circuit containing registers and combinational logic. The CP-graph changes its state in discrete time-steps $t=0,1,2,\dots$, the transition $t \rightarrow t+1$ can be computed by a labelling procedure described below.

At time t any storage node contains definit information and the edges are unlabelled. Starting out from this situation, any edge leading out of a storage node is labelled with the information contained in the storage node. The edges leading out of the IOP node are labelled with the input information sent by the I/O processor. If all the edges leading to a PE-node are labelled, then the operation assigned to the PE-node is carried out, and the edges leading out of the node are labelled with the results. This computation rule is applied until all PE-operations are executed and all edges are labelled. (It is not difficult to prove that this situation is reached in any case, if the CP-graph is constructed considering the rules described above.) At time t+1 any storage node inputs the label of the edge leading to it, storage nodes without input edge keep their information. At the same time all labels are deleted in the graph, and the transition t ⟶ t+1 is finished.

A CP-graph can be regarded as a special algorithm, which computes some results from some initial data. Namely, initial data are

- the initial information in the storage nodes at time 0,
- the input data sent by the I/O processor in each step;

results are

- the output data sent to the I/O processor in each step,
- the final information in the storage nodes after the last step.

3. BASIC PROCESSING ELEMENTS

In this section we give a set of PE-nodes which seems to be sufficient to construct CP-graphs to any kind of computational problems. We have developed cellular PEs to each of these PE-nodes, in most cases several versions with different area-time parameters are available to a given PE-node. This background of cellular PEs ensures a cellular program implementation to any CP-graph constructed from the elements given below.

Fixed-point arithmetics. Input and output operands are k-digit signed binary numbers in two's complement code. The PE-nodes and their notations are the following:

adder: 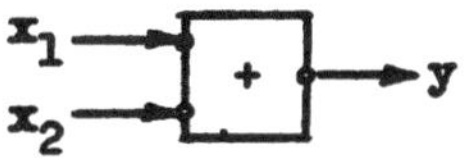

$(y=x_1+x_2)$

complementer: 

$(y=-x)$

multiplier: 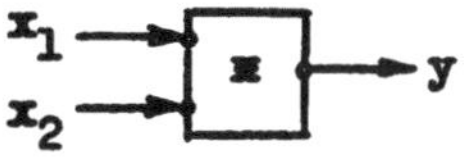

$(y=x_1 \cdot x_2)$

divider: 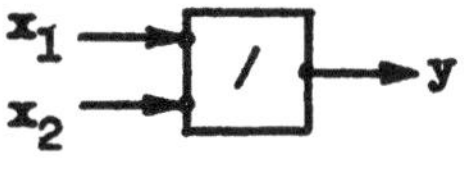

$(y=x_1/x_2)$

Substraction can be realized by the interconnection of a complementer and an adder, but a direct cellular substractor can be constructed, too.

Fixed-point cellular PEs are described in [5],[6],[9].

Floating-point arithmetics. Input and output operands are floating-point numbers. Similar types of PEs are developed as in the case of fixed-point operations (in notation the word "float" may be used for distinction). The PEs do not normalize the result, therefore we need an additional PE for normalization:

Floating-point cellular PEs are discussed in [10].

Boolean operations. Each cell of a cellprocessor can be regarded as a Boolean processor, so Boolean PEs can be realized obviously using only one cell. Notation:

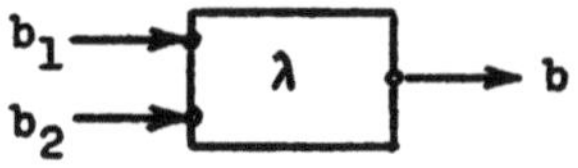

where b_1, b_2 and b are Boolean variables, λ is a Boolean operation and $b = b_1 \lambda b_2$.

Comparator. A fixed-point number is compared with zero, the result of the comparison is a Boolean variable b. Notation:

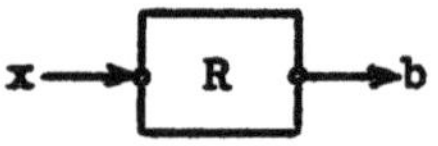

where $R \in \{=, \neq, <, \leq, >, \geq\}$ and b=1 if xR0 holds, b=0 otherwise. Floating-point comparison can be made by comparing the mantissa as a fixed-point number. Comparator PEs are discussed in [5].

Selector. The result of a comparison is generally used for the control of an algorithm. In a cellular algorithm such a control is performed by applying selector PEs, which receive three input operands: two numbers, x_1 and x_2, and a control bit b. The output is one of the input numbers chosen by b. Notation:

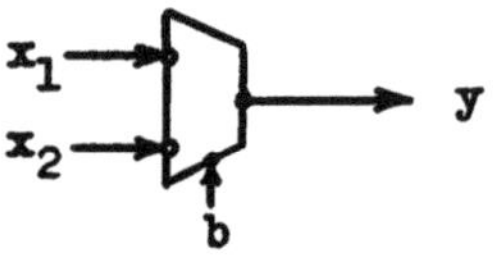

where $y=x_1$ if b=1, $y=x_2$ otherwise.

The above set of PE-nodes is complete in the sense that any computational problem seems to be solvable by a CP-graph built from that set. On the other hand, it proved necessary to construct further PEs for some frequently occuring functions as sorting [8], associative search [7], etc. Although these latter PEs can be derived from the above surveyed ones, such a derivation would result in a redundant implementation.

4. AN EXAMPLE: DIGITAL FILTERING

In digital filtering the formula

$$y_n = \sum_{i=o}^{m-1} d_i x_{n-i} - \sum_{i=1}^{m} b_i y_{n-i}$$

is usually applied producing an output sequence of signals $y_1, y_2, \ldots$ from an input sequence $x_1, x_2, \ldots$. Since all the signals fall into a definit interval, they can be coded as fixed-point numbers.

Fig.6 shows a CP-graph of a digital filter with m=3. This graph is very similar to a special-purpose hardware realization.

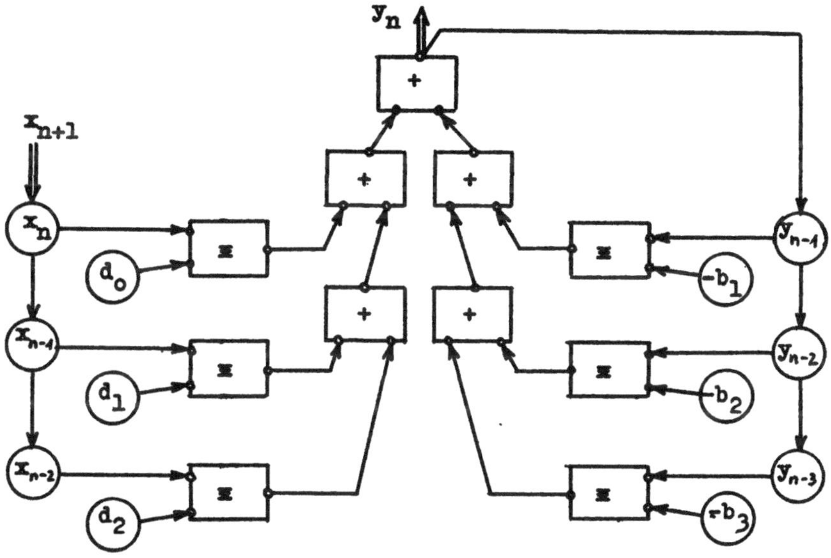

Fig.6. CP-graph for a 3-stage recursive filter.

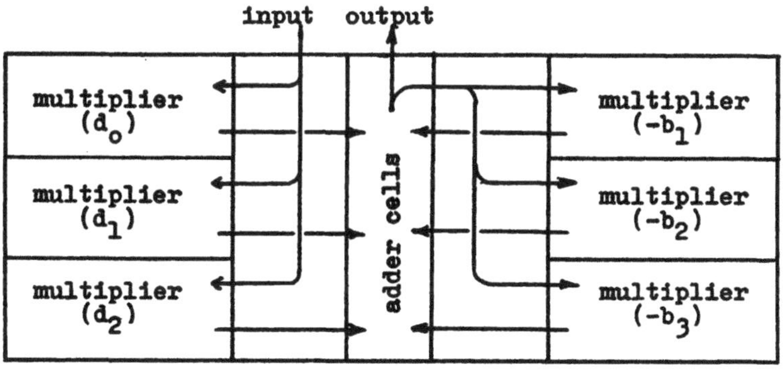

Fig.7. Sketch of a cellular program for a 3-stage recursive filter.

A cellular program implementation of the CP-graph is sketched in fig.7. Multipliers work with bit-serial input and output, each multiplier occupies an area of 3 by k/2 cells, if four-bit cells and k-bit numbers are supposed. The constants d_i and b_i are stored in the multiplier cells bit by bit. The storage nodes of the CP-graph are replaced by data channels ensuring the delays required for synchronization. For details see [11].

5. DESIGN METHODOLOGY

CP-graphs make it possible to decompose the hard problem of cellular programming into easier subproblems. The following strategy may be applied in constructing complex cellular programs.

The initial step is to design a CP-graph for the given computational problem. The CP-graph can be built from the PE-nodes surveyed in section 3, but in special application fields (e.g. in data processing [8]) another set of PE-nodes may be used.

The next step is to assign concrete cellular PEs to the PE-nodes of the CP-graph. Here we should take into consideration the size of the cellular field. (High-speed PEs can be chosen with great area requirements if the cellular field is large enough, low-speed ones with less area requirements otherwise [5]). The interconnected PEs should have the same succeeding time in any case (see section 1).

Synchronization in the CP-graph. A weight is ordered to each edge of the graph meaning necessary delays in the data channels. These weights can be computed from the concrete delay parameters of the concrete cellular PEs. In the course of synchronization, storage nodes are eliminated in the CP-graph.[1)]

Embedding into the cellular field. The PEs are placed into the cellular field so that any edge of weight w is mapped to a data channel of length w. "Length" here does not mean a real distance, because the teletransport ensures zero-length data channels between farther cells.

The sketched design methodology sets up a challange for the future, to produce an automatic translation (performed by conventional software) from the CP-graph to a cellular program. Since CP-graphs can be described in some data-flow like language, an automatic translation would offer the possibility of programming a cellprocessor in a high-level language. The most problematical point of automatic translation is the embedding into the cellular field, because it seems that an area-optimal embedding cannot be done in general without the help of human intelligence.

---

1) It is possible that the synchronization of a cyclic CP-graph can be made only by increasing the succeeding time of the PEs. If this increase is great, the cellular program becomes inefficient.

REFERENCES

[1] Ch.E. Leiserson and J.B. Saxe: Optimizing synchronous systems. Proceedings of the 22nd Annual Symposium on Foundations of Computer Science, IEEE Computer Society, October 1981, pp. 23-36.

[2] T. Legendi: Cellprocessors in computer architecture. Computational Linguistics and Computer Languages XI, 1977, pp. 147-167.

[3] T. Legendi: Cellular algorithms and their verification. Proceedings of the Conference CONPAR'81, Lecture Notes in Computer Science 111, pp. 169-188.

[4] J. Tóth: A 64 K cell LSI based cellprocessor prototype. Proceedings of the Workshop "Parallel Processing by Cellular Automata and Arrays", 1984. (this conference proceedings)

[5] E. Katona: Bit-parallel processing with bit-sequential and bit-parallel data handling in cellular space. Computers and Artificial Intelligence 2 (1983) 315-329.

[6] E. Katona: Binary addition and multiplication in cellular space. Acta Cybernetica, v.5, 1982, pp. 457-464.

[7] E. Katona: A software realization of an associative storage in a cellprocessor and its applications. Proceedings of the Braunschweig Cellular Meeting 1981, T.U. Braunschweig, pp. 47-56.

[8] E. Katona: The application of cellprocessors in conventional data processing. Proceedings of the Third Hungarian Computer Science Conference, Publishing House of the Hungarian Academy of Sciences, Budapest, 1981, pp. 295-306.

[9] F. Csordás: The implementation of Monte Carlo methods in a cell-processor. Thesis, József A. University, Szeged, 1983 (in Hungarian).

[10] G. Fodor: Floating-point arithmetics in cellular space. Thesis, József A. University, Szeged, 1984 (in Hungarian).

[11] Z. Szántai: Application of cellprocessors for digital filtering. Thesis, József A. University, Szeged, 1984. (in Hungarian)

[12] H.T. Kung: The structure of parallel algorithms. Advances in Computers, v.19, Academic Press, New York, 1980, pp. 65-112. (ed.: M.C. Yovits)

[13] U. Bernutat-Buchmann, D. Rudolph, K.-H. Schlosser (ed): Parallel Computing I. A bibliography of 5000 items. Rechenzentrum der Ruhr-Universität Bochum, 1983.

FROM CELLULAR ALGORITHMS TO CELLULAR PROGRAMS

Peter Köles[1/]

1. Introduction

In the last years many cellular /systolic/ algorithms have been published [1, 2, 4-7, 12]. They are exact descriptions of parallel algorithms, which are suitable to implement on cellular processor /cellprocessor/ architectures. However many further information, and maybe a bit transformation, are necessary so that these cellular algorithms should become fast and powerful cellular programs on a concrete cellprocessor architecture which has concrete abilities and parameters.

In Szeged concrete cellprocessor architectures have been planned [10, 11, 13] therefore the development of cellular programs has also started using the parameters of cellprocessor architectures which have been fixed till now.

The paper shows the parts of cellular programs and shows the way from cellular algorithms to cellular programs showing the important details of cellular programs. Some concrete examples are given as an illustration in the last part of the paper.

2. Cellular processor architectures

This point shows the broad outlines of cellprocessor architectures [8] and gives the function of the parts of the cellprocessor architectures. Each cellprocessor architecture consists of three main parts /Fig. 1./ such as the cellular space, the cellular space control unit /CCU/ and the input-output processor /IOP/.

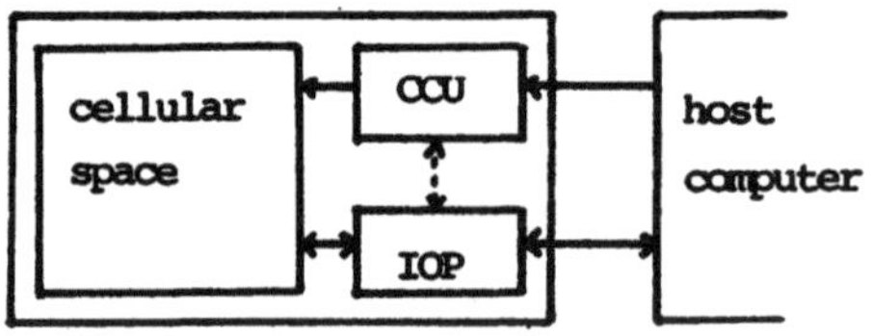

Fig. 1. Cellular processor architecture

1/ Research Group on Automata Theory of Hungarian Academy of Sciences
H-6720 Szeged, Somogyi u. 7., Hungary

The main characteristics of this system are the following:

(i) The cellular space is a rectangular cell-matrix bounded by I/O edge buffers. In the cellular net the von Neumann neighbourhood is assumed. Microcells are applied having maximum 16 states.

(ii) The cells receive commands /microinstructions/ from CCU, so an arbitrary local transition function may be realized by the execution of microinstructions. This means that from step to step different transition functions may be used.

(iii) Each cell has an internal state. The cells with different internal states work according to different transition functions, selected by their internal states at the same time.

CCU not only sends the microprograms of transition functions but it controls the cellular space, too. IOP controls input-output transmission between the cellular space and the host computer and performs the necessary conversion of data. Generally CCU and IOP work in parallel. They are synchronized in an implicit way.

The concrete architectures are different in the type of the cells, the size of the cellular space, the input-output organization and so on.

3. Cellular algorithms

A cellular algorithm is a special parallel algorithm based on local operations. As a consequence of the construction of the cellprocessor architecture cellular algorithms are expected to have the following properties:

- data must be split into bits or groups of bits because of the limited processing and storing capacity of one cell
- operations must be executed on the data meanwhile they are moving in the cellular space
- operations executed in the same time must not be of great many varieties /relative homogeneity/
- real operations must be performed in each place and time /high efficiency/.

A great number of cellular algorithms have been developed in Szeged [1-2, 4-5, 12]. They represent many different application areas such as arithmetic and matrix operations, picture processing, syntax analysis and so on. Many algorithms have a few different versions. These versions are important from the view-point of cellular programs because the most suitable version can be chosen for a concrete cellular program. The available cellular algorithms constitute a good basis of the cellular program development.

4. Cellular programs

A cellular program is the realization of a cellular algorithm on a concrete cellprocessor architecture. Therefore the cellular program must contain all the information which are necessary for the work of the cellprocessor. The most important details of them are the initial external and internal states of cells, the time variable assignment of transition functions during the processing and the input-output management. The assignment of transition functions is realized as the main part of the program of CCU. The input-output management and the loading of the initial states are the parts of the program of IOP.

4.1. Choosing a suitable cellular algorithms

For a concrete cellular program we must choose a cellular algorithm from the available set. We must consider the concrete size of the cellular space, the type of the cells and the execution time of the cellular algorithm. We can choose from bit-parallel and bit-serial cellular algorithms. The execution time of the bit-parallel algorithms independent of the length of the data, but they need larger cellular space while bit--serial algorithms usually need smaller cellular space but more time /depending on the data length/ for the computation [2].

If the available set of cellular algorithms does contain no algorithm suitable in all respect, we must transform a similar algorithm into a suitable one. In most cases it means the reduction in the state-bits number of a cell to one bit, because the concrete cellprocessors have two--state cells. There exists a general procedure [3] which describes such a transformation, that is, this transformation always can be performed. However in many cases the use of some special transformations and some sophisticated tricks instead of the general method results in a better transformed algorithm. A difficult task is to make the most of the special abilities of the concrete cellprocessors. These abilities are different, but each ability is very useful, so the algorithms will be faster and more puwerful. For example special abilities of [11]: two layers of cellular space, dynamic internal bits /the dynamic internal state bit may be changed by the transition functions during the computation/.

4.2. Microprogram of transition functions

The kernel of the CCU program is the assignment of transition functions. A transition function is described by a sequence of microinstructions. It is the most important requirement that the microprogram should be optimal because the sequences of transition functions are sent to the cellular space in every step. It is not sure, that the optimal microprogram may be created from the simplest transition function . A local operation may be described by different transition functions. The transition function with the shortest microprogram is used.

4.3. Loading the initial states

A cellular algorithm describes the initial /external and internal/ states of cells and the inhomogeneity of processing elements. /A processing element is a group of cells performing certain elementary function as addition, multiplication [2]./ The cell-programmer must decide how to load the initial external states /they may be loaded by shifting from outside or directly by special transition functions/ and how to divide the real internal states among the cells and among the different processing elements. It also must be decided which internal bits are to be handled dynamically.

4.4. Input-output management

The organization of input-output needs more attention. The data transmission must be planned not only between the cellular space and the IOP but it must be planned between the IOP and the host computer, too.

Data are usually converted the required format by IOP /input-output data format i.e. geometrical ordering of the data bits/ but data may be converted in the cellular space and, in a few simple cases, by the host computer, too. The cellprogrammer decides where the conversion is performed, that is where the fastest and the simplest it is. He must create the suitable program for the format conversion but in most cases there exist standard conversion routines.

5. Examples

In this point the most interesting phases of cellular program development are shown by a simple example. Let us start from the cellular algorithm of the bit matrix multiplication [1]. The task is to compute the matrix C of size nxn by the following formula:

$$c_{ij} = \sum_{t=1}^{n} a_{it} b_{tj} \qquad i,j = 1,\ldots,n$$

where the matrices A and B of size nxn are given.

The multiplication is the logical "and" operation, the addition is the logical "or" operation. The sketch of the cellular algorithm is given in Fig. 2.

The transition function /it is described by the so-called microconfiguration term [9]/:

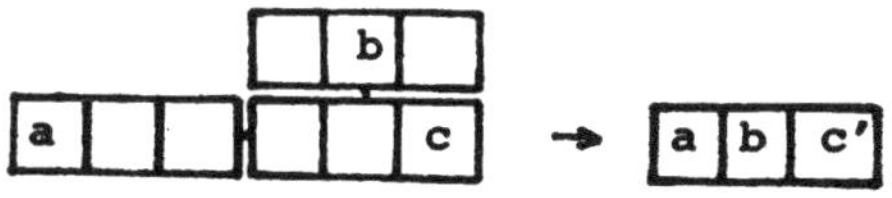

where $c' = c \vee (a \wedge b)$

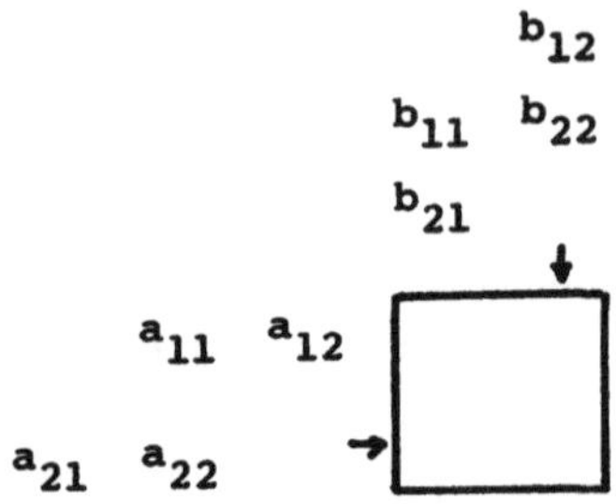

Fig. 2. Bit matrix multiplication /n=2/

As we can see, the algorithm needs 3-bit cells. The concrete cell-processors have one bit cells. Therefore we must transform the basic algorithm into two-state cells. Instead of one 3-bit cell we take four 1-bit cells in a square form. It is called a block. In this case two steps are needed instead of one step of the basic algorithm. The transition functions of the transformed algorithm are the following:

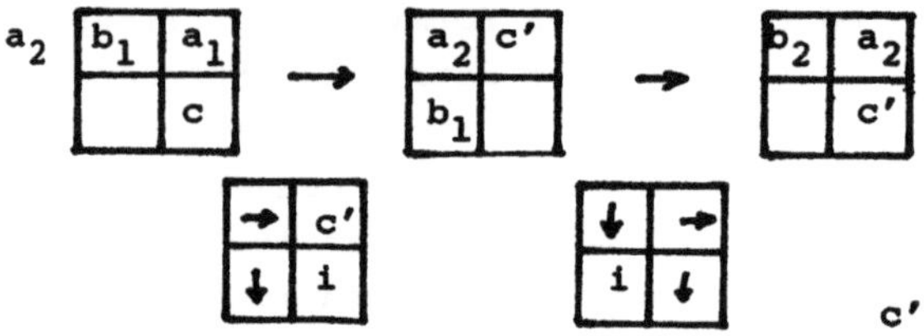

$c' = c \vee (a_1 \wedge b_1)$

Shifting transition functions are marked by arrows, "i" means the identical transition function.

This transformed algorithm may be applied on the cellular processor model [10] and the cellprocessor based on traditional components [13] , but these cellprocessors have a very small cellular space. Therefore the input matrices must be cut into "stripes", as it is shown in Fig. 3.

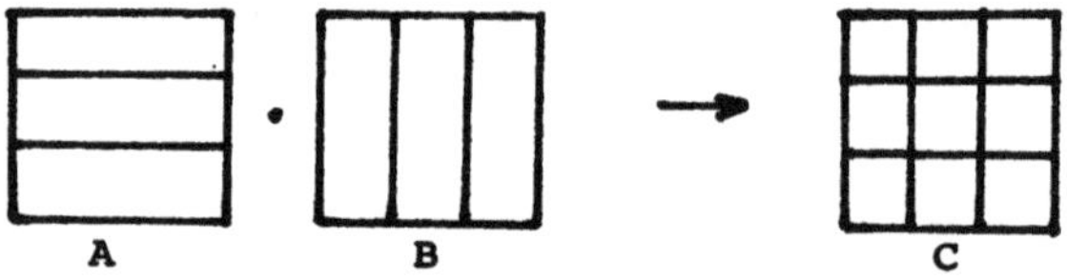

Fig. 3. Partitioned matrix multiplication

In one time period one part of matrix C is computed by the cellular space. The greatest problem is the input-output management. How to organize so that data transmission should be minimum and the cellular space should not wait for data. In the case of cellprocessor model data management was planned by the following:

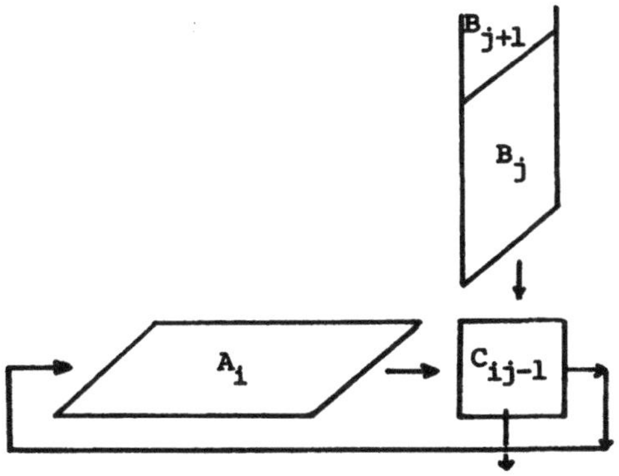

Fig. 4. Input-output management in the case of cellprocessor model

The cellprocessor model has some FIFO-s for input-output. Input FIFO-s are loaded by the IOP and the other end of the input FIFO-s are in connection with the cellular space. On the other hand, the end of the FIFO-s loaded by IOP may be connected with another edge of the cellular space, as it is shown in Fig. 4. Therefore A_i goes around, meanwhile it is multiplicated with the stripes of matrix B. After a "round" one part of the matrix C is available. In this way matrix A is transferred only once.

In the case of the cellprocessor prototype the size of the blocks may be reduced using a dynamic internal bit. Only one one-bit cell is used instead of one 3-bit cell of the basic algorithm. Let us consider the transition function of the basic algorithm. We can see, that the bit c of the cells is not necessary for their neighbours. Therefore it may be computed by an internal bit. By thinning the data transfer /Fig.5./ we get a new transformed algorithm.

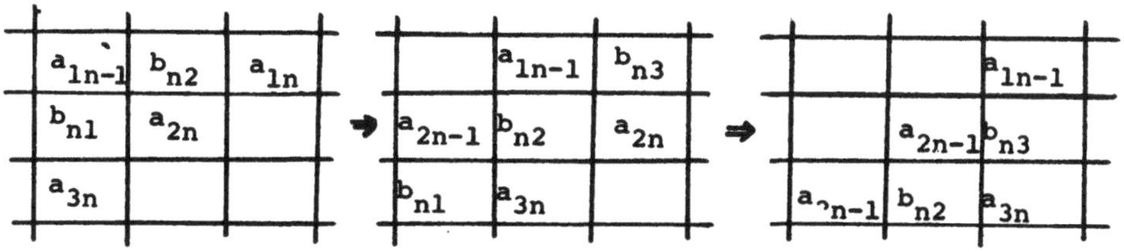

Fig. 5. Data transfer in thinning form

The transition function is the following /double frame marks the dynamic internal bit/:

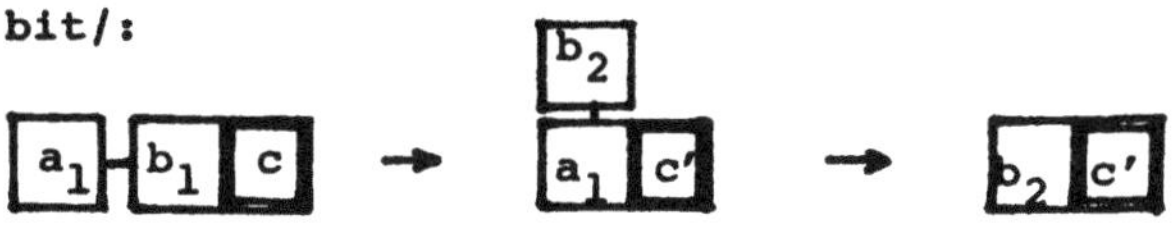

where $c' = c \vee (a_1 \wedge b_1)$

Using this transformed algorithm we can develop a simple and, for this very reason, fast cellular program for the cellprocessor prototype.

Conclusions

The basis of the cellular program development is a wide set of cellular algorithms. Cellular algorithms created in Szeged till now ensure such a wide set. Using certain standard methods combined with some tricks transformed algorithms can be created easily which are suitable for direct development of cellular programs considering the former requirements. In this way fast and powerful cellular programs can be developed which can satisfy the demands of users utilizing the special abilities of the different cellprocessors.

References

[1] Katona, E.: Cellular algorithms for binary matrix operations. Proceedings of the conference CONPAR'81, Nürnberg 1981, Lecture Notes in Computer Science, 111,Springer, 1981, pp. 203-216.

[2] Katona, E.: Bit-parallel processing with bit-sequential and bit--parallel data handling in cellular space. Computers and Artificial Intelligence, Vol. 2, No. 4, 1983, pp. 315-329.

[3] Katona, E.: Programming methods of two-state cellular spaces. Manuscript /in Hungarian/

[4] Köles, P.: Digital picture processing by cellprocessor. Computers and Artificial Intelligence, Vol. 2, No.4, 1983, pp. 331-339.

[5] Köles, P.: The realization of a simplified assembler in cellular space. Proc. of the conf. Parallel Computing'83, Elsevier Science Publisher B.V. /North-Holland/, 1984, pp. 437-442.

[6] Kung, H.T.: Let's design algorithms for VLSI Systems.Report No. CMU-CS-79-151, Carnegie-Mellon University, Pittsburgh, 1979.

[7] Kung, H.T.: Why systolic architectures?, Computer, Jan. 1982, pp. 37-46.

[8] Legendi, T.: Cellprocessors in computer architecture. Computational Linguistics and Computer Languages 11, 1977, pp. 147-167.

[9] Legendi, T.: A 2D transition function definition language for a subsystem of the CELLAS cellular processor simulation language.Computational Linguistics and Computer Languages 13, 1979, pp. 169-194.

[10] Tóth, J.: Design of a cellprocessor model.presented at I. Int. Workshop on Parallel Processing by Cellular Automata, Berlin, 1982.

[11] Tóth, J.: A 64 Kcell LSI based cellprocessor prototype /SPP/. Presented at II. Int. Workshop on Parallel Processing by Cellular Automata and Arrays, Berlin, 1984.

[12] Vollmar, R.-Legendi, T. /eds/: Cellprocessors and Cellgorithms. Report on a common Research Project, Informatik-Skripten No.1. T.U. Braunschweig, 1981.

[13] Zsótér, A.: Cellprocessor based on traditional components. Presented at II. Int. Workshop on Parallel Processing by Cellular Automata and Arrays, Berlin, 1984.

CELLULAR PROGRAM FOR THE SOBEL TRANSFORMATION ON A TRADITIONAL COMPONENT BASED CELLPROCESSOR

István Kovács[1/]

1. Introduction

After constructing algorithms for an ideal cellprocessor recently we have developed algorithms for concrete architectures. Hardware engineers have designed some rather different cellprocessor architectures, and some of them are under realization.

We can read in [4] about a "small" cellprocessor based on traditional components. Here the word "small" means that this architecture consists of a few hundred 2 state cells.

In this lecture we should like to show that such "small" cellprocessors can be used efficiently to perform relatively complex algorithms.

Picture processing is one of the most important applications of cellprocessors, and Sobel transformation is a wellknown and efficient method for edge enhancement. Sobel transformation is defined in the following way. We fit a 3x3 window to the picture as shown in Fig. 1.

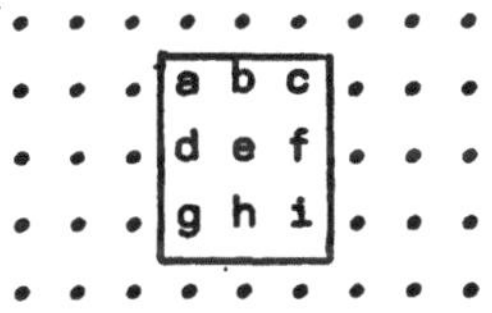

Fig. 1. A window on the picture

where a,b,...,i are pixels.

We define $S_X=(c+2f+i)-(a+2d+g)$ and

$S_Y=(g+2h+i)-(a+2b+c)$.

The new value of the pixel e will be $S = |S_X| + |S_Y|$.

2. About the cellprocessor

The cellprocessor /Fig. 2., [4]/ contains 2 layers. The upper layer can only perform shift functions so it executes data conversions. The real cellular space is the lower layer.

There are 15x16 cells /15 rows, 16 columns/.

Cells have 2 /external/ states,

32 internal states and

the inhomogeneity in time may be of 16 types.

Each cell receives a 4 bit information from the CCU in each step

1/Research Group on Automata Theory of Hungarian Academy of Sciences
H-6720 Szeged, Somogyi u. 7. Hungary

which is used for choosing one from transition functions. This 4 bit information may be different in each step, and in this way "inhomogeneity in time" is ensured.

The cells are connected according to the von Neumann neighbourhood.

The input comes at the upper edge of the space, and the output leaves at the lower edge of the space. There is an 8 line buffer connected to the upper layer. The input is sent to the space from the buffer and the buffer receives the output.

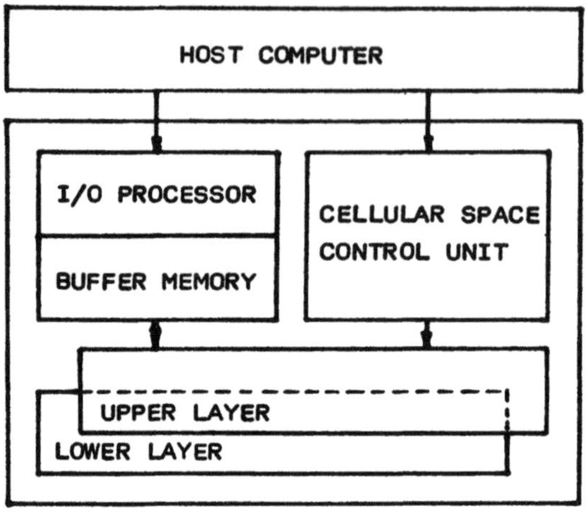

Fig. 2. The cellprocessor

3. Organization of data flow

One bufferline can store one line of the picture. At the same time 3 picture-lines are stored in the buffer. The pixels of these lines will be the input, and the new values of the middle line pixels will be the output to be stored in the 4th line of the buffer. Calculation of the new value of a pixel requires the reading of the pixel and its 8 neighbours twice. Let E be the j-th pixel of the i-th picture line, and let A,B,...,I be its neighbours.

To calculate the new value of E the pixels are sent to the space in the following sequence:

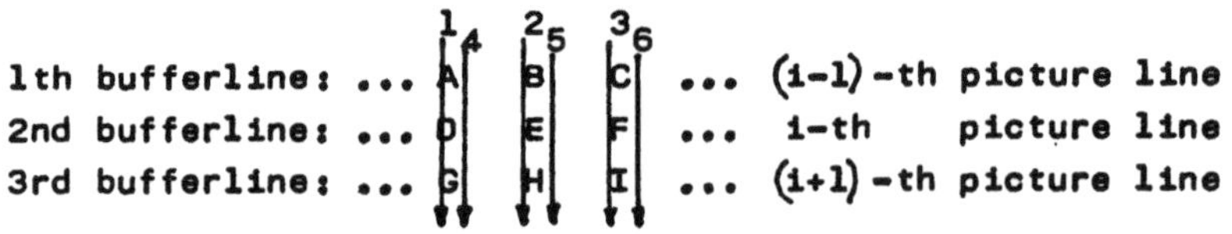

that is the ADGBEHCFI pixel-sequence is sent to the space twice. First $S_X(E)$ and second $S_Y(E)$ are calculated. After calculating the new values of i-th line pixels the (i-1)-th line is deleted, and the (i+2)-th line will be sent to the buffer from the picture to calculate the new values of (i+1)-th line pixels.

The six bit pixels are sent to the space in slanting and thinning form, such as

· · · · · ·
A_6 · D_4 · G_2 ·
· · · · ·
A_5 · D_3 · G_1
· · · ·
A_4 · D_2 ·
· · ·
A_3 · D_1
· ·
A_2 ·
·
A_1

Here we can see that a new pixel enters the space in every 4th step.

The output form is the following:

· · · · · A_1
· · · · A_2
· · · A_3
· · A_4
· A_5
A_6

4. About the algorithm /Partitioning of the cellular space/

To calculate S_X's the pixels of the window are multiplied by the mask:

$$\begin{matrix} -1 & 0 & 1 \\ -2 & 0 & 2 \\ -1 & 0 & 1 \end{matrix}$$

and similarly to calculate S_Y's the following mask is used:

$$\begin{matrix} -1 & -2 & -1 \\ 0 & 0 & 0 \\ 1 & 2 & 1 \end{matrix} .$$

These multiplications are performed in the following sequence:

- First the pixels are multiplied by 0 or by 1 or by 2.
- Then they are multiplied by -1 or by 1.

/Negative numbers are represented in two's complement code./

Then these 9 numbers have to be added and then S_X and S_Y are ready.

The next operation is the computing of absolute value (CAV) and then $|S_X|$ and $|S_Y|$ are added.

Generally this sum consists of 9 bits, so the result has to be "rounded" in order to get a 6 bit pixel.

So we get a partitioning of the cellular space as shown in Fig. 3.

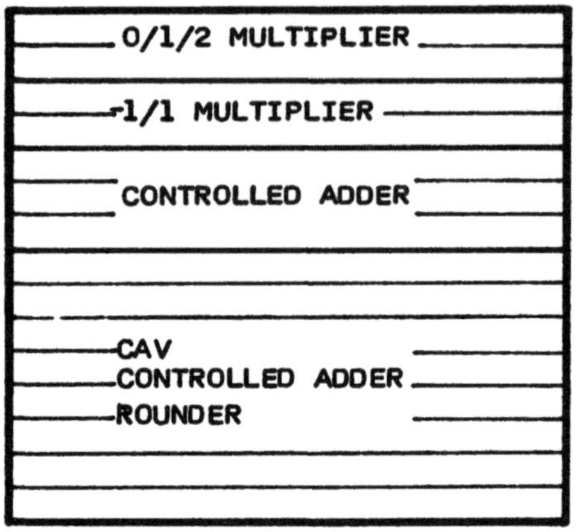

Fig. 3. Partitioning of the cellular space

Data flow is unchanged till the end of the 3rd partition. The change occurs in the 4th partition as we shall see in the next point.

Hence the first 3 operations work in a 4 step cycle. Each part of the cellular space is controlled by different signals, which in most cases are generated in the rightmost column.

5. Elements of the algorithm

In this paragraph A, B denote the input-numbers of a partition and, accordingly, A_i, B_i denote the i-th bit of the number A, B, respectively. The output numbers of a partition are denoted by α, β and the i-th bits by α_i, β_i, respectively. The signals concerning A will be V_A and W_A.

5.1 0/1/2 multiplier

2 lines of the space and 2 signals /V and W/ are used.

If the pixel is multiplied by 0 1 2 the signals will be:

	0	1	2
V:	0	0	1
W:	0	1	0 .

We show 4 steps to trace data flow and the changing of transition function assignments, which are repeated cyclically /see Fig. 4./.

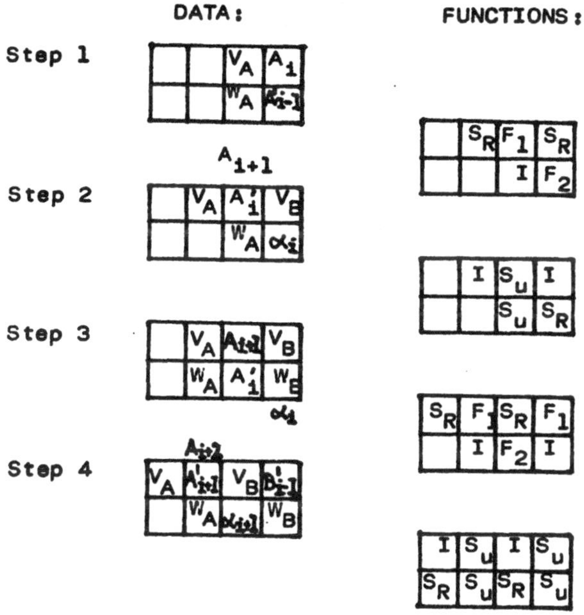

Fig. 4. Data flow and the changing of function assignments /0/1/2 multiplier/

Notations of functions are the following:

/The given function concernes the cell denoted [ ] ./

I: "identical" function 

S_u: "shift from up" function 

S_R: "shift from right" function 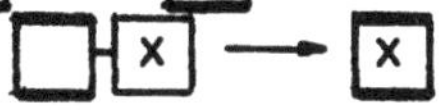

F_1: 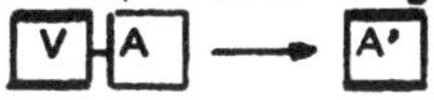 $A' = V \cdot A$

F_2: 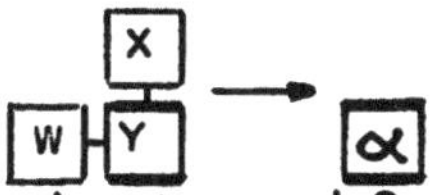 $\alpha = W \cdot X \oplus Y$

/ $\oplus$ denotes mod 2 addition/

Signals get their value from following sequences cyclically:

V: 010000010000101000
W: 101000101101000101

5.2 -1/1 multiplier

Multiplying a number by -1 is equivalent to the complementing of the number. This operation can be performed if we let the bits unchanged upto the first bit of value 1 from right and the other bits are negated.

2 rows of the space and 2 signals /V and W/ are used.

$$V = \begin{cases} 1 & \text{if the number has to be complemented} \\ 0 & \text{otherwise} \end{cases}$$

The signal of W is 0 at the beginning and it is changed into 1 when it reaches the first 1 bit of the number.

This part of the space also works in 4 step cycle as shown in Fig. 5.

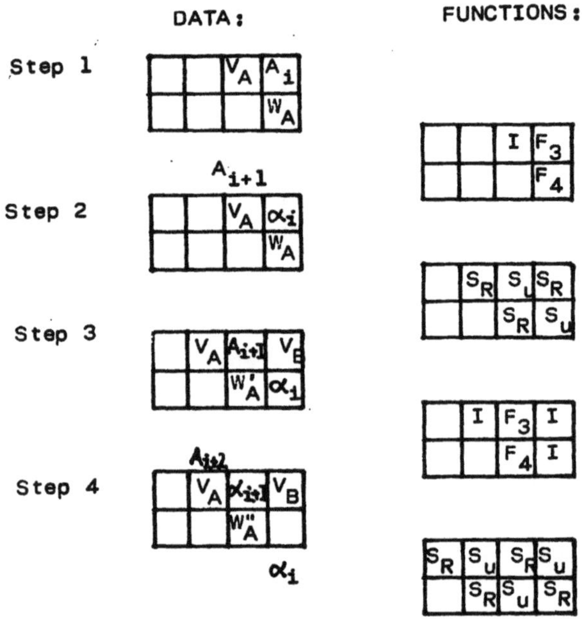

Fig. 5. Data flow and the changing of function assignments /-1/1 multiplier/

Function-notations:

F_3:

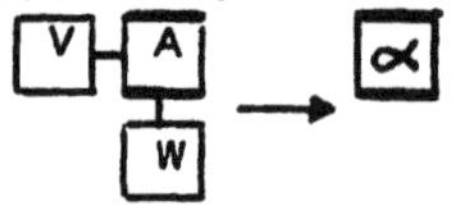

$$\alpha = (\neg A \wedge V \wedge W) \vee (A \wedge \neg V \wedge W) \vee (A \wedge V \wedge \neg W) \vee (A \wedge \neg V \wedge \neg W)$$

F_4: 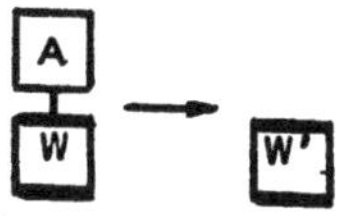

$W' = A \vee W$

The signal of V gets its value from the following sequence cyclically.

V: 11100000100100100

5.3 Controlled adder

To compute S_X or S_Y 9 numbers must be added. This operation is performed in 3 rows of the space with a signal V.

$$V = \begin{cases} 1, & \text{if the result is read from the upper row, and this row must be deleted /that is if the 9 numbers have already been added/.} \\ 0 & \text{otherwise} \end{cases}$$

The signal gets its value from the following sequence cyclically:

V: 000000001 .

This partition also works in 4 step cycle as we can see in Fig. 6. /C_i denotes carry bit and S_i denotes sum bit./

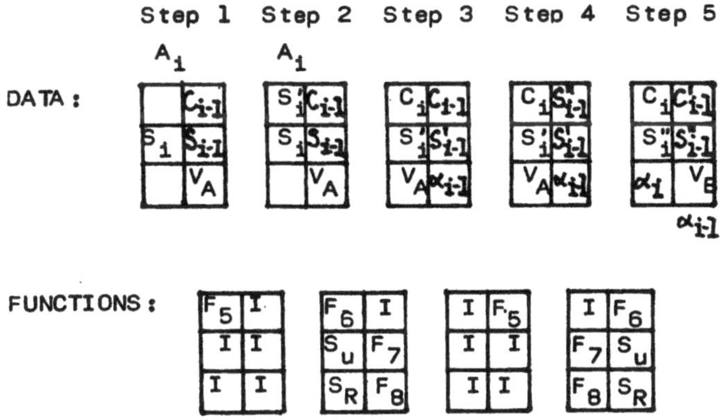

Fig. 6. Data flow and the changing of function assignments /controlled adder/

Function-notations:

F_5: 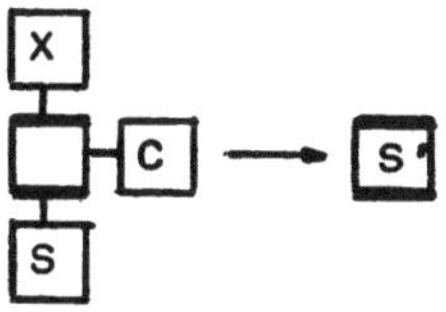

$S' = S \oplus C \oplus X$

F_6:

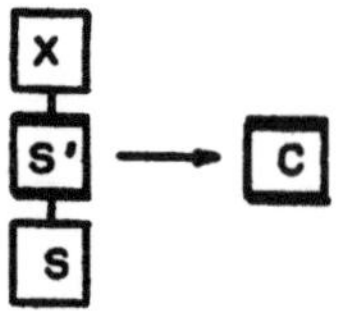

$C=(X+S+(X \oplus S \oplus S'))/2$

F_7:

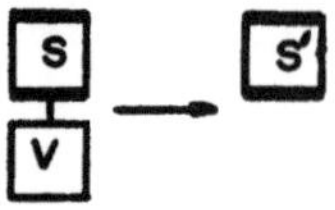

$S'=\neg V \wedge S$

F_8:

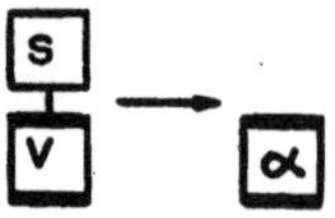

$\alpha = V \wedge S$

The sum of S_X and S_Y consists of 8 bits and a sign bit. After every 36 steps a new sum will be resulted, since the adder has to read 9 numbers which come every 4th step.

5.4 The absolute value of S_X's and S_Y's are computed in the other 8 rows of the space, then these $|S_X|$, $|S_Y|$ pairs are added and at the end the result is "rounded".

5.4.1 First the sum is transformed into a straight form using a signal of $V(=1)$ in 2 rows as we can see in Fig. 7.

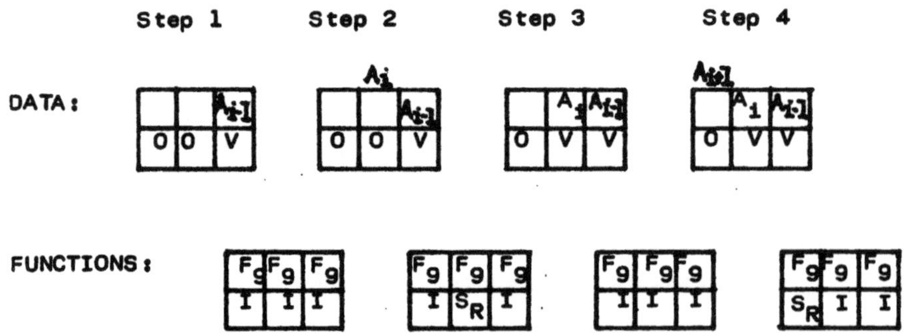

Fig. 7. Data flow and the changing of function assignment /Transformation into a straight form/

Function notation:

F_9:

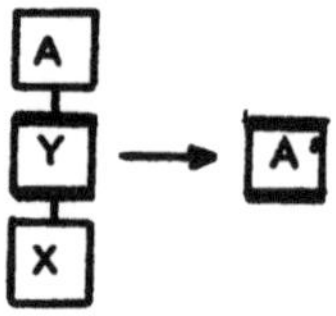

$$A' = \begin{cases} A, & \text{if } X=0 \\ Y, & \text{if } X=V(=1) \end{cases}$$

So the sum is received in the 10th row in the next form:

$SA_8A_7 \ldots A_1$, where S is the sign.

$X_8X_7 \ldots X_1$

We can see the number X in the 11th row which is

$$X = \begin{cases} 0, & \text{if A is a } S_X \text{ sum} \\ S_X, & \text{if A is a } S_Y \text{ sum} \end{cases}$$

5.4.2 Computing the absolute value of a negative number is equivalent to the complementing of the number. This operation is performed in a way different from the one used previously.
If S = 1 /the number is negative/, every bit is negated and then 1 /in our case it will be S/ is added to the number.
If S = 0, the number is unchanged.

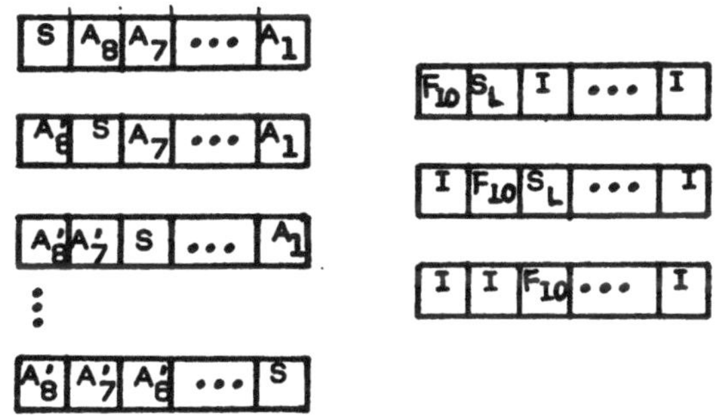

Fig. 8. Data flow and the changing of function assignments /Computing of the absolute value/

Function notations:

F_{10}: 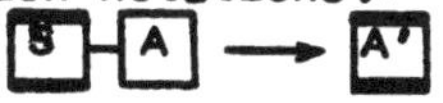 $A' = (\neg S \wedge A) \vee (S \wedge \neg A)$

S_L: "shift from left" function 

Then an addition is performed. On one hand to finish computing the absolute value the S bit has to be added to number A, and on the other hand the A+X addition has to be performed. We can consider the bit of S as the first carry bit. This addition is similar to the previous /controlled addition/, so here steps are not detailed.

5.4.3 If the result of the addition is $|S_X|$ then it is left unchanged, and if it is $|S_X| + |S_Y|$ then it has to be read out. In the latter case we have to "round" the result, since it consists of 9 bits. We cut the first 3 bits and if one of the first 3 bits is 1, the pixel will be

111111, otherwise the lower 6 bits will remain unchanged. These two operations can be performed with 3 signals /V, W and W/ as we can see in Fig. 9, where

$V''' = 0$

$$W = \begin{cases} 0, & \text{if the } A = |S_X| + |S_Y| \text{ number has to be read out and delete this number} \\ 1, & \text{if the } A = |S_X| \text{ number has to remained unchanged.} \end{cases}$$

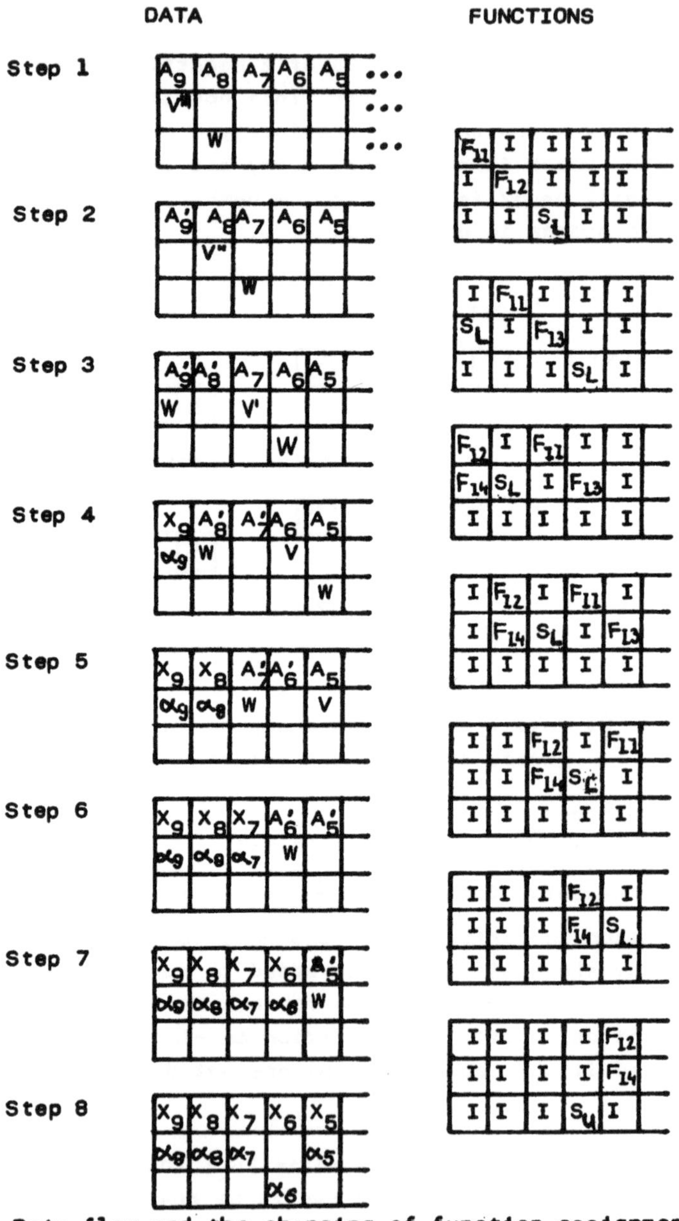

Fig. 9. Data flow and the changing of function assignments /Rounding/

Function notations:

F_{11}: 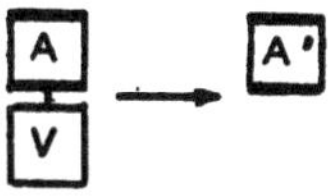

$A' = V \wedge A$

F_{12}: 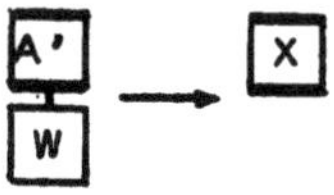

$X = W \wedge A'$

F_{13}: 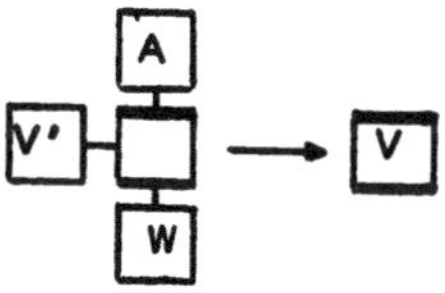

$V = (A \vee V') \wedge \neg W$

F_{14}: 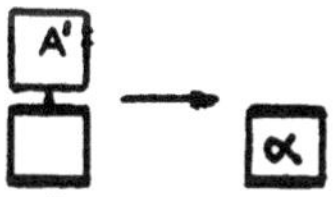

$\alpha = \neg W \wedge A'$

6. Conclusion

Let us estimate the time complexity of the Sobel transformation. Suppose the picture consists of 512 x 512 pixels and a step of the cellspace lasts 100 nsec. In this case the processing of the picture needs less then 2 seconds.

References

[1] Duda, R. O. and Hart, P. E.: Pattern Classification and Scene Analysis /A Wiley-Interscience Publication 1973/

[2] Hirsch, Á.: Realization of combined arithmetical operations in cellular space. Thesis, JATE University Szeged 1983. /In Hungarian/

[3] Vollmar, R. - Legendi, T. /eds/: Cellprocessors and Cellgorithms. Report on a common research project, Informatik-Skripten No. 1. T. U. Braunschweig, 1981

[4] Zsótér A.: Cellprocessor based on traditional components, Presented at II. Int. Workshop on Parallel Processing by Cellular Automata and Arrays, Berlin 1984

CEPROL - A Cellular Programming Language

Friedhelm Seutter
Institut für Theoretische und Praktische Informatik
Technische Universität Braunschweig
Federal Republic of Germany

Abstract

A programming language is presented which offers means for programming and controlling pipeline algorithms running on cellular automata.

0. Introduction

The performance of a computer system has increased by magnitudes during the last four decades. This increase essentially refers to faster logical circuits and their minimization. For a long time the classical von Neumann-architecture and a sequential processing of programs has been kept. Adding I/O-processors or other special-purpose processors working parallel to the central processing unit does not change this basic conception.

One way for a performance increase of computer systems is to parallelize the computing power. But only by comming up (V)LSI-technology and dramatically dropping hardware costs this way became possible. Two well-known types of parallel computers are the array-processor and the multiprocessor (Stone, 1975). According to the classification of Flynn (Flynn, 1966) the first belongs to the single-instruction multiple-data stream (SIMD) computers and the latter to the multiple-instruction multiple-data stream (MIMD) computers. The von Neumann-architecture belongs to the single-instruction single-data stream (SISD) computers. Another technic for increasing the performance is pipelining (Chen, 1975).

A model of a parallel computer is the cellular automaton. A cellular automaton consists of a finite number of Moore-type automata being fixed to the gridpoints of a finite-dimensional space. The automata are connected by a homogeneous pattern. The states of the automata are changed synchronously according to a given function (e. g. Vollmar, 1979).

A direct realization of cellular automata is not possible today, because of still existing hardware problems. But some restricted virtual cellular nets are developped (Legendi, 1981a) and already realized (Höllerer et al., 1981). They are restricted in neighbourhood and in number of the states of a cell. Such a cellular automaton will not be operated as a stand-alone computer, but as a special-purpose processor in a universal computer system (Legendi, 1976). This kind of operation is quite appropriate, because so just the parallelizable tasks may be served by the cellular automaton and all the services offered by the host system may be used. Realized cellular automata therefore are well qualified for handling pipeline-processes.

There is a similar idea in the direct realization of algorithms on VLSI-chips (Foster, Kung, 1980). The following demands are required on these algorithms: Implementation by only a few different types of simple cells; data and control-flow is simple and regular, so that the cells can be connected by a network of simple and regular interconnections; and the algorithm uses pipelining. Such algorithms are called "systolic". These special-purpose processors are attached to a universal computer system as peripheral devices. A cellular automaton also is appropriate for processing systolic algorithms and in this way it may be operated as a programmable special-purpose processor.

Programming cellular automata is done essentially by transition tables, wheather realized (Höllerer et al., 1981) or simulated (Höllerer, 1980; Vollmar, 1973). This method is oriented to the internal representation of cellular algorithms by tables and not to the requirements of their applications. Therefore a cellular programming language is introduced. It is supposed to program and control cellular automata processing pipeline-algorithms.

1. Cellular programming and pipelining

A cellular automaton which is attached to a computer system has to be programmed and controlled by the host. There are instructions necessary for initialization, I/O of data and execution of global transitions. This global control mechanism may be done by special procedure calls out of conventional programming languages.

The main task is the programming of the cellular automaton. Therefore a local program is assigned to every cell. A global behaviour of the cellular automaton is defined by arranging the local programs topologically and chronologically. It is also possible to program just parts of a cellular net and to compose them on a higher level.

This structure can also be found in pipeline-processes. The solution of a problem is divided in functional units and a result is obtained by passing through these units. Therefore the transformation of pipeline-processes to cellular algorithms is really simple.

2. Summary of the language

The programming language CEPROL is defined as a PASCAL extension (Jensen, Wirth, 1978). In its added instruction set there are means for programming and controlling cellular algorithms which essentially simulate pipeline-processes. This recourse to a conventional programming language is appropriate, because of the connection of the cellular automaton to a host system.

A cellular algorithm is controlled out of a PASCAL statement part by so-called net-procedures. They are concerned to initialization of the cellular net, execution of global transitions and net I/O. But the essential part of the language extension is made up by means for programming (i. e. defining) a cellular net.

The basic objects for programming cellular nets are local functions and data channels. A local function is a local program for one cell to modify data stored in this cell, a data channel is a path in the cellular net to move data through the net. The net is programmed by arranging these objects topologically and chronologically. It is possible to program subnets -named modules- to support the modular structure of a cellular algorithm. The cellular net may be composed of local functions, data channels and modules.

A CEPROL-program is setup in the following way: It begins with a program heading. There the program is assigned a name and its formal parameters (names of input and output files) are listed. The following program body consists of a net definition section and an execution section. The net definition section again is divided in a number definition part, a nettype definition part and a net definition part. In a net definition local functions, modules and data channels are defined and arranged in the net. Syntactically a net definition consists of a number definition part, nettype definition part, local function definition part, module definition part, data channel definition part and setting part. The execution section consists essentially of a PASCAL block. But the language is extended by predefined net-procedures for programming net control and net I/O.

A survey of the program parts is given in figure 1 by a 'skeleton' of a CEPROL program and in figure 2 by syntax diagrams.

```
CEPROG ceprogid ( . . . );                        } program heading

    NUMBER . . . ;                                  ┐
    NETTYPE . . . ;                                 │
                                                    │
    NET netid1 (NETIO num : IOSTREAM);     ┐        │
      [                                    │        │
      [                                    │ net    │
    NET netid2 (NETIO num : IOSTREAM);     │ definition
      [                                    │ section│
      [                                    │        │
    .                                      │        │
    .                                      │        │
    .                                      ┘        │ program body
    LABEL      . . . ;  ┐                  ┐        │
    CONST      . . . ;  │                  │        │
    TYPE       . . . ;  │ definition part  │        │
    VAR        . . . ;  │                  │ execution
    PROCEDURE  . . . ;  │                  │ section│
    FUNCTION   . . . ;  ┘                  │        │
                                           │        │
    BEGIN               ┐                  │        │
      [                 │ statement part   │        │
      [                 │                  │        │
    END.                ┘                  ┘        ┘
```

Fig. 1: 'Skeleton' of a CEPROL program

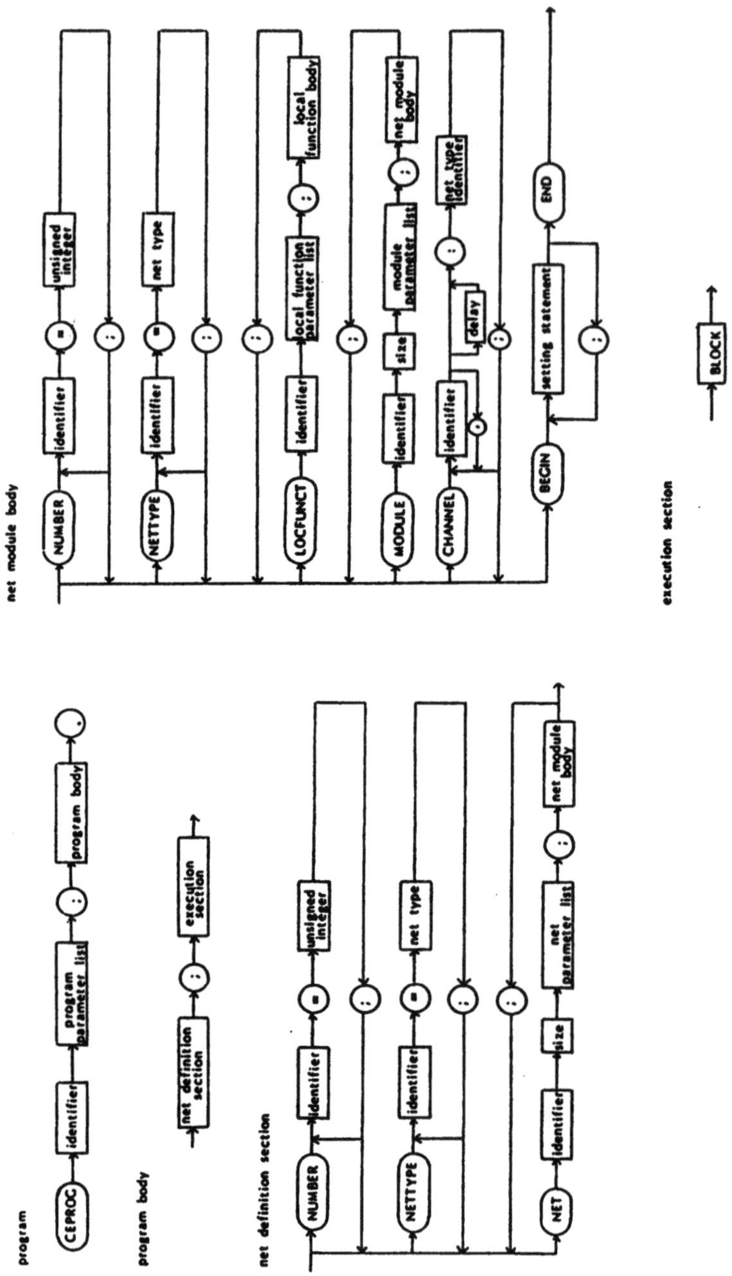

Fig. 2: Syntax diagrams of a CEPROL program

The programming language is block structured (figure 3). A block is composed of a heading and a body. The outermost block is the whole program. Blocks in the net definition section are a net definition, a local function definition and a module definition; in the execution section the PASCAL block structure holds. Net definition section and execution section are not a block.

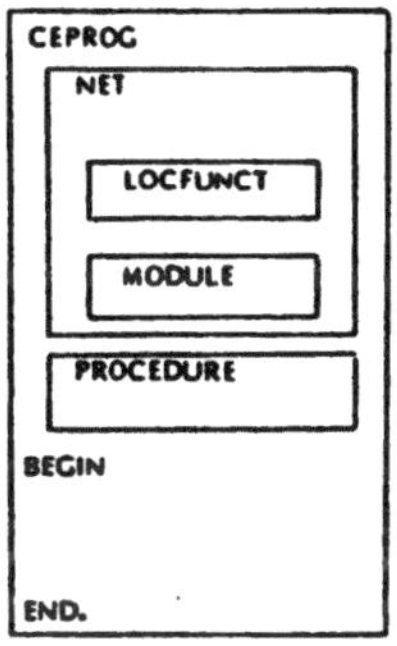

Fig. 3: Block structure

3. An example

As a programming example a bit-sequential adder for any number of summands with output of intermediate sums (Legendi, 1981b) is given. Because of possibly occurring carries, the summands have to enter the adder in a slanting format. The sums then leave the adder in this slanting format, too (figure 4).

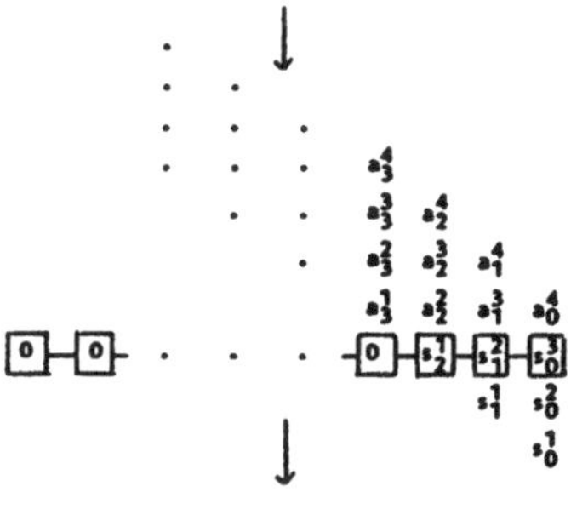

Fig. 4: Schema of a bit-sequential adder for any number of summands $a^k=a^k_n \ldots a^k_1 a^k_0$. $s^k=s^k_n \ldots s^k_1 s^k_0$ are intermediate sums. $a^k_i, s^k_i \in \{0,1\}$, $0 \leq i \leq n$, $1 \leq k \leq m$, $n,m \in N$.

Let the summands be $a^k = a_n^k \ldots a_1^k a_0^k$ and the intermediate sums $s^k = s_n^k \ldots s_1^k s_0^k$, $a_i^k, s_i^k \in \{0,1\}$, $0 \le i \le n$, $1 \le k \le m$, $n,m \in N$. The adding cells have to execute the functions

$$s_i^k = s_i^{k-1} \oplus c_{i-1}^{k-1} \oplus a_i^k$$

and

$$c_i^k = \begin{cases} 1 \text{ if } s_i^{k-1} + c_{i-1}^{k-1} + a_i^k > 1 \\ 0 \text{ otherwise} \end{cases}$$

where not defined variables have the value 0, $\oplus$ is modulus-2-addition and c_i^k are carries.

The numbers are loaded/unloaded bit-parallel into/out of the cellular net. Therefore the transformation in/from slanting format also is done in the net. To gain these transformations a delaying subnet is set before and behind the adder. There the moving of the binary values in the data channels is delayed correspondingly. Figure 5 shows a sketch of a so programmed cellular net. Then the CEPROL program of this adder is given.

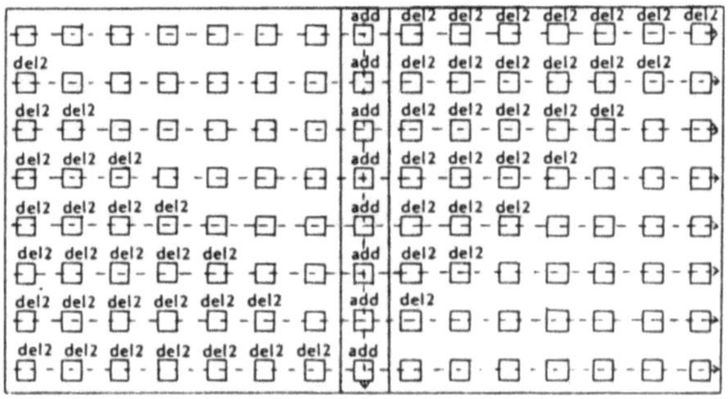

Fig. 5: Sketch of a cellular net which is programmed as an adder with output of intermediate sums.

```
CEPROG adderiso (infile,outfile);
{Pipelineadder for any number of summands (integers) with output of
 intermediate sums}

NUMBER n = 16;                      {Number of binary positions of a
                                     summand}

NET addnetiso[n,2*n-1] (NETIO num : IOCHANNEL);
{Adder with delaying subnets for transforming the summands in/from
 slanting format. Net input and output via data channel "num".}

   LOCFUNCT delay2 (INVAR ain : bit; OUTVAR aout : bit);
   {Delaying function with delay 2}

      LOCVAR store : bit;
      BEGIN
         store := ain;
         aout  := store
      END; {end "delay2"}

   MODULE sum[n] (CHANNEL num : bit);
   {Bit sequential adder for n bit numbers. The binary values are
    entering and leaving the module via the data channel "num" which
    is set outside the module.}

      LOCFUNCT add (INVAR ain,cin : bit; OUTVAR sout,cout : bit);
      {Adder of binary values, where "ain", "cin" are input variables,
       "sout", "cout" are output variables for summands, sums and
       carries, respectivly.}

         LOCVAR s : bit;          {binary store for sums}
         BEGIN
            s,sout := ain + s + cin;                        {sum}
            cout   := ain * s OR ain * cin OR s * cin       {carry}
         END; {end "add"}

      CHANNEL carry : bit;        {data channel for carry}

      BEGIN
         carry ON CELL[1] TO CELL[n];
         FORALL i:=1 TO n DO
            add(num,carry,num,carry) ON CELL[i]
      END; {end "sum"}

   CHANNEL num : bit;     {binary data channel for summands and sums}

   BEGIN
      FORALL i:=1 TO n DO
         num ON CELL[i,1] TO CELL[i,2*n-1];
      FORALL i:=1 TO n-1, j:=1 TO i DO
         delay2 (num,num) ON CELL[i+1,j], CELL[n-i,n+j];
      sum (num) ON CELL[1..n,n]
   END; {end "addnetiso"}
```

```
VAR countin, countout,                {counter for summands and sums}
    i      : integer;                  {additional variable}
    buffer: ARRAY[1..n] OF bit;        {net input/net output buffer}
    infile,outfile : FILE OF integer; {input/output file}

BEGIN
{Each step a summand is written into the net and a sum is read out of
 it. There is a delay of 3n-2 steps between net input and output.}

   loadnet(addnetiso,buffer);       {initialization of the cellular
                                     net and attaching of the input
                                     and output buffer}
   reset(infile); rewrite(outfile);
   countin:=0; countout:=0;

   WHILE NOT eof(infile) DO
   {next summand}
   BEGIN
      intbin(infile^,buffer); get(infile); {Reading next summand from
                                             input file and converting
                                             in an array of bits deno-
                                             tation}
      netin(buffer);                        {net input}
      step(1);                              {transition}
      countin:=countin+1;

      IF countin >= 3*n-2
      THEN
      {sums are ready for output}
      BEGIN
         netout(buffer);                    {net output}
         binint(outfile^,buffer); put(outfile);
                                            {converting in an integer
                                             denotation and writing a
                                             sum on output file}
         countout:=countout+1
      END
   END;
   {input file is empty}

   IF countin < 3*n-2
   THEN
   {delay of a summand in the cellular net is greater than the number
    of summands}
      step(3*n-3-countin);                  {transitions}

   FOR i:=1 TO countin-countout DO
   {reading the remaining sums out of the net}
   BEGIN
      step(1);                              {transition}
      netout(buffer);                       {net output}
      binint(outfile^,buffer); put(outfile){converting in an integer
                                             denotation and output of
                                             a sum}
   END
END.
```

References

Chen, Tien Chi (1975): Overlap and Pipeline Processing.
In: Stone, H. S. (ed.): Introduction to Computer Architecture. Chicago, 375 - 431

Flynn, M. J. (1966): Very High-Speed Computing Systems. Proceedings of the IEEE, Vol. 54, No. 12, 1901 - 1909

Foster, M. J., Kung, H. T. (1980): The Design of Special-Purpose VLSI Chips. Computer, 1, 26 - 40

Höllerer, W. O. (1980): SILICEA - A Simulator for 'realizable' Cellular Automata. Computational Linguistics and Computer Languages 14, 95 - 135

Höllerer, W. O., Henkel, V., Hammer, C. (1981): BVCP - The Brunswick Virtual Cellular Processor.
In: Vollmar, R., Legendi, T. (eds.): Cellprocessors and Cellgorithms. TU Braunschweig, Informatik-Skripten, 1, Braunschweig, 255 - 276

Jensen, K., Wirth, N. (1978): PASCAL, User Manual and Report. New York

Legendi, T. (1976): Cellprocessors in Computer Architecture. Computational Linguistics and Computer Languages, 11, 147 - 167

Legendi, T. (1981a): Programming of a Cellular Processor Model.
In: Vollmar, R., Legendi, T. (eds.): Cellprocessors and Cellgorithms. TU Braunschweig, Informatik-Skripten, 1, 52 - 61

Legendi, T. (1981b): Cellular Algorithms and their Verification.
In: Vollmar, R., Legendi, T. (eds.): Cellprocessors and Cellgorithms. TU Braunschweig, Informatik-Skripten, 1, 24 - 51

Stone, H. S. (1975): Parallel Computers.
In: Stone, H. S. (ed.): Introduction to Computer Architecture. Chicago, 318 - 374

Vollmar, R. (1973): Über einen Interpretierer zur Simulation zellularer Automaten. Angewandte Informatik, 6, 249 - 256

Vollmar, R. (1979): Algorithmen in Zellularautomaten. Stuttgart

A 64K CELL LSI BASED CELLPROCESSOR PROTOTYPE /SPP/

József Tóth 1)

The cellprocessor concept

From the point of view of realization many types of parallel computers constitute a compromise between processing elements, memory, switching network and hardware technology. One of them is the cellprocessor [7] which is an MIMD computer and is derived from von Neumann's cellular automaton. Figure 1. shows the structure of a cellprocessor which has the following features:

1. A cellprocessor is not a stand-alone computer but is a part of a traditional computing system. It works under the control of a host computer and executes those tasks which fit to its architecture and capability

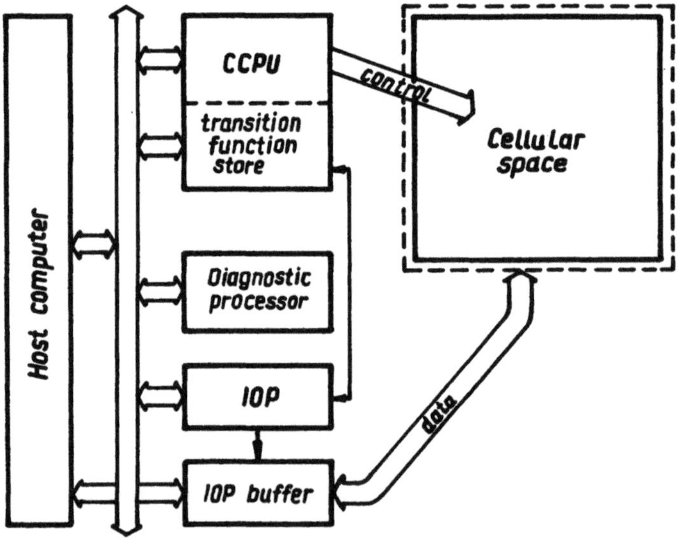

Fig. 1. The cellprocessor structure

1) Attila József University, Computer Science Department
Somogyi u. 7. H-672o Szeged, Hungary

2. A cellprocessor has - beside hardware homogeneity - an inhomogeneous cellular space both in time and in space. As a result of this inhomogeneity PE-s are constructed by several neighbouring cells, and the cellprocessor is totally software reconfigurable. On the other hand inhomogeneity ensures programmability of the space
3. Cells are very simple instruction executing units and - beside the local connections - are connected to a common broadcasting bus. This bus is driven by a central unit /CCPU/. Transition functions are stored in the CCPU in the form of terms or microinstructions and are executed by serial broadcasting.

The SPP chip

The basic component of the SPP is a full custom designed LSI chip which consists of an 8x8 array of cells and a switching network. A 6µm silicon gate NMOS technology is used and the chip is packed into a 28 pin DIL package. Estimated power consumption is about 1W from a single 5V power supply.

The basic cell

The cells are mesh-connected and have Neumann neighbourhood. Figure 2. is the block diagram of the basic cell. The S bit is the outer state of the cell /two states/, the T bit is the so-called passive layer storage bit /see later/, the M bit is an accumulator bit which the next state of the cell is computed in. The E bit stores a partial result of the comparison between the contents of the bus and either the states of the neighbouring cells or the internal state of the cell. A six bit internal state storage ensures the space inhomogeneity. The bits of the internal state storage can be used as a scratch-pad memory when computing the next state /dynamic cell/.

The basic cell has six different 16 bit microinstructions and the working out of a total /inhomogeneous/ step of the space needs 1o...4o microinstructions depending on the complexity of the transition function.

Switching network

An 8x8 array of cells in a chip is called a block. Blocks are not connected to form a Neumann space but they are connected by a switching network.

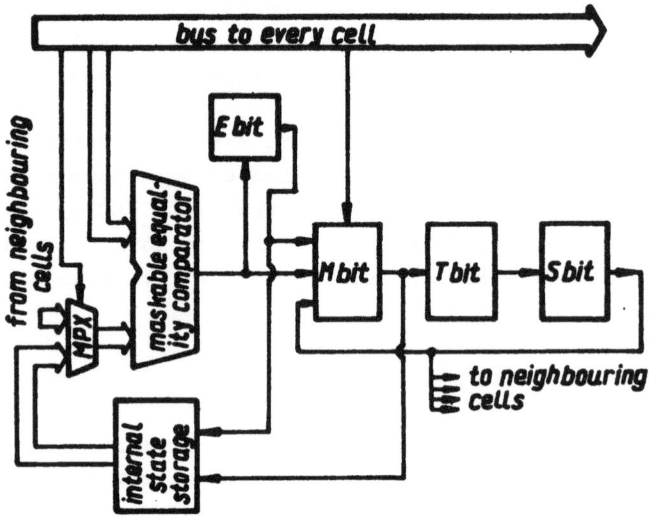

Fig. 2. Block diagram of the basic cell

The switching network consists of two parts, namely a fixed part /thick arrows on Fig. 3./ and a programmable part /striped areas on Fig. 3./. The programmable part in a chip can connect the edges of the block in a chip and the fixed part. A memory in the programmable part stores the information of connections. Usually two neighbouring blocks are connected forming a Neumann subspace but it is possible to jump over a block. A way which connects non-neighbouring blocks is called a teletransport path. Teletransport paths are very useful because remote /software/ PE-s can be connected and the input-output processor can communicate with internal PE-s.

Architecture of the SPP

Cellular space

Cellular space can be seen as a 256x256 array of Neumann neighbourhood cells of Figure 2. but in certain places a byte-of-cells-in-line is connected to /a/ non-neighbouring byte/s/-of-cells-in-line. To be more precise edge cells of remote blocks can be connected by teletransport paths.

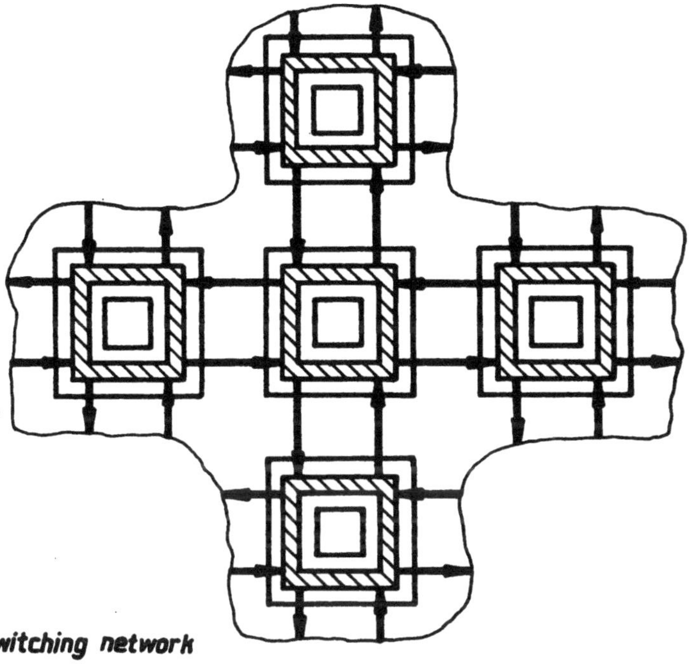

Fig.3. Switching network

Teletransport paths may have a limited number of crossovers and they must be defined before the run of the SPP.

In order to eliminate the time of teletransport the cellular space has two layers. The two layers have common internal state information and work on a chopped mode. While in one of the layers /active layer/ the next state is computed, in the other /passive layer/ the teletransport and interchip data exchange is carried out [Legendi].

An 8x4 array of blocks constitutes a so-called field. Fields have different cellprograms and these cellprograms run paralelly.

Every block within a field has an internal state bit /per chip internal state/.

These features increase the space inhomogeneity and make the cellular programming easier.

Cellular central processing unit /CCPU/

CCPU is a vector machine. An element of the vector is called a CCPU-element. The CCPU-elements are assigned to the fields, have different microprograms and work paralelly. The CCPU-elements are synchronized at the end of every total cellular space step.

Figure 4. shows the architecture of a CCPU-element. Every element has a serial program stored in its memory. Beside that it has a micro-program counter, a microinstruction decoder and register, a cycle counter, a stack and a comparing and synchronization logic.

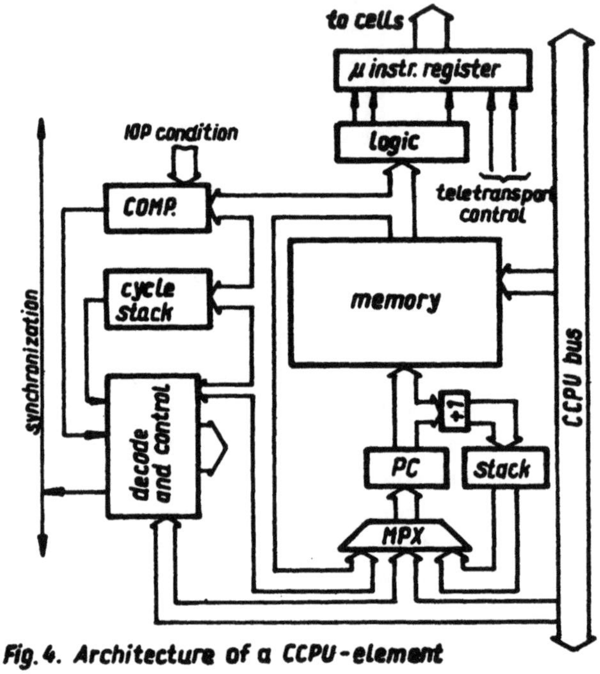

Fig. 4. Architecture of a CCPU-element

Microinstructions of a CCPU-element can be divided into two groups. The group of cellular-space-microinstructions /transition function is coded into a sequence of microinstructions/ are broadcasted to the space while CCPU-program-control-microinstructions /subroutine call, looping, conditional jump, etc./ are processed by the CCPU-element. Microinstructions of two different kinds of groups can be paralelly executed.

Input-output processor

Figure 5. is the block diagram of the input-output processor.

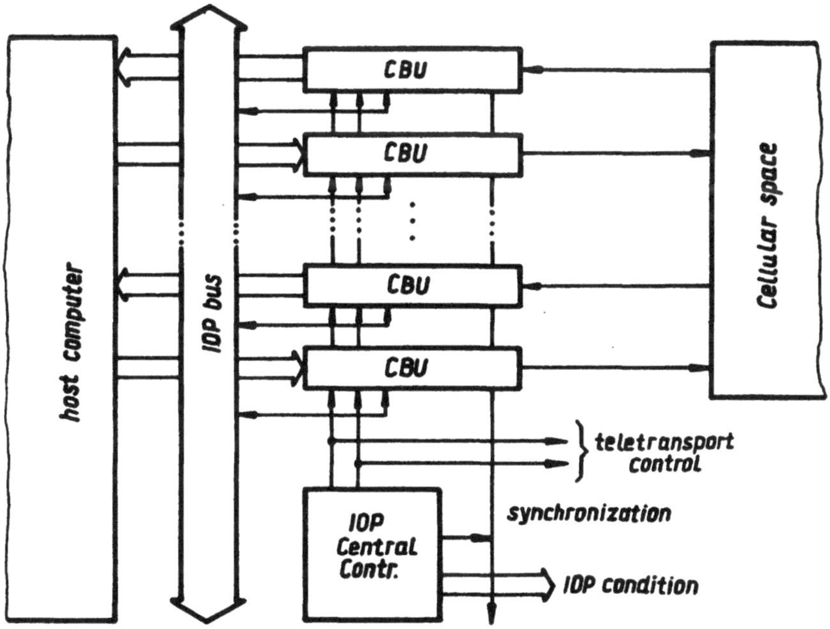

Fig.5. Block diagram of the IOP

The input-output processor has two tasks. The first one is the control of the teletransport process in the passive layer. The second one is data handling between the host computer and cellular space. The teletransport can be controlled by special microinstructions of the SPP chip therefore the IOP Central Controller has a microprogram which is similar to the program of a CCPU-element. This program is derived from the definition of teletransport paths.

The Converter-Buffer Units /CBU-s/ control the data flow. CBU-s work paralel to each other because they have their own microprograms. A CBU's input or output is connected to the input or output of the host computer or to a teletransport path in the space. As data flow through a CBU, the CBU converts /into a slanting or bit-serial format etc./ and buffers them.

The CCPU and IOP are synchronized on hardware level.

SP/P microassembly language

```
PATH  ct,cr1,cr2,...
PATH ...                 } definition of teletransport paths (ct: coordinate of
  :                        the transmitting byte-of-cells-in-line; cri: co-
                           ordinate of the receiving byte(s)-of-cells-in-line
END
CBU      n1,n2,...
  : } µinstrs            } definition of the CBU µprograms
CBU ...                    (ni: number of CBU)
  :
END
LAYER
FIELD  c1,c2,...
PHASE
  : } µinstrs            } definition of CCPU program
FIELD ...                  (ci: coordinate of field)
  :
END
ENDPROG
```

Fig. 6. Simplified structure of an SP/P program

The SP/P language is a microassembly language and used to describe SPP cellular programs. Using this language every architectural capability of the SPP can be exploited. Figure 6. shows a simplified SP/P program. This program consists of three parts.

At first the teletransport paths are defined. A teletransport path is an one-directed way with one transmitter and several receivers. The compiler examines the structure of crossovers and the length of the teletransport paths, then generates the content of the memories which are placed in the switching network as well as generates the program of the IOP Central Controller. Control information of neighbouring chip data exchange is automatically generated.

The CBU microprograms define the data handling. Data are converted to special format by bit manipulation. This manipulation is described by a serial microprogram.

The part which describes the control of the cellular space begins with assign directives. These directives assign a part of the cellular space, and the microprogram placed after them controls the assigned part only. Cellular space control microprograms are also serial ones.

Literature

[1] Batcher, K.E.:Design of a Massively Parallel Processor, IEEE Trans. on Comp.,Vol. C-29,No.9,September 198o,pp. 836-84o.

[2] Katona, E.:SP/P Microassembly Language /CCPU component/,/manuscript in Hungarian/,Research Group on Automata Theory of the Hungarian Academy of Sciences, Szeged, 1984.

[3] Kung, H.T. and Leiserson, C.E.:Systolic Arrays /for VLSI/, in Duff, I.S. and Stewart, G.W. /eds./:Sparse Matrix Proceedings 1978, /Society for Industrial and Applied Mathematics, 1979/

[4] Legendi, T.:Cellular Algorithms and their Verification, "CONPAR '81" Nürnberg, June 1o-12, 1981.

[5] Legendi, T.:Implementation and Application of Cellular Processors, Proceedings of "Parallel Computing '83", West Berlin, 1983.

[6] Tóth, J. and Legendi, T.:Documentation of the SPP chip, /manuscript in Hungarian/, Attila József University, Computer Science Dept. Szeged, 1984.

[7] Vollmar, R. and Legendi, T. /eds./:Cellprocessors and Cellgorithms /Report on a Common Research Project/, Technische Universität Braunschweig,1981.

CELLPROCESSORS BASED ON TRADITIONAL COMPONENTS

Antal Zsótér[1/]

One way of implementation of cellprocessors is to make special integrated circuits, containing tens or hundreds of cells, and then to build cellular space or cellprocessor from them. A circuit containing 64 sorting cells has already been made, and two circuits containing 64 universal 2-state and 16-state cells are under development [1]. These cellprocessors are called custom-designed LSI-based cellprocessors. The custom-designed LSI circuit based universal cells are so-called quasi-cells /fig. 1/. These quasi-cells don't contain their transition functions, they are sent out in series to each cell by a so-called cellular space control unit through command lines /fig. 1/. This centralized sequential transition function processing drastically reduces the size of cells, thus making it possible to integrate tens or hundreds of cells on one chip, and making it more flexible by variable transition functions [2].

Another way of implementation of cellprocessors is to build cells or cellprocessor from standard /catalog/ circuits. Such cellprocessors are called traditional component cellprocessors. A cellprocessor-model containing standard component quasi-cells has already been made [3]. Another cellprocessor - which is the topic of this article - is under development. This cellprocessor contains so-called stored-function cells. Stored-function cells contain their transition function in a transition function memory of their own /fig. 1/.

A standard component cell needs several /2-4/ packages, so in the same volume we can build much smaller cellular space from catalog-circuits than from custom-designed LSI circuits, where one circiut contains tens or hundreds of cells. However standard component cells are faster than cells based on custom-designed LSI circuits, because standard component stored-function cells contain their transition functions in a transition function memory of their own, and these functions need not be sent out in series.

Generally we can say that cellprocessors implemented from custom-designed LSI circuits are better:

- We can solve tasks requiring large cellular spaces.
- Their processing rate is high /100-1000 MOPS/ according to many parallel operations in the large cellular space /many cells work in parallel/.

But there are many applications requiring not too large cellular spaces,

1/ Attila József University H-6720 Szeged,Somogyi u. 7. Hungary

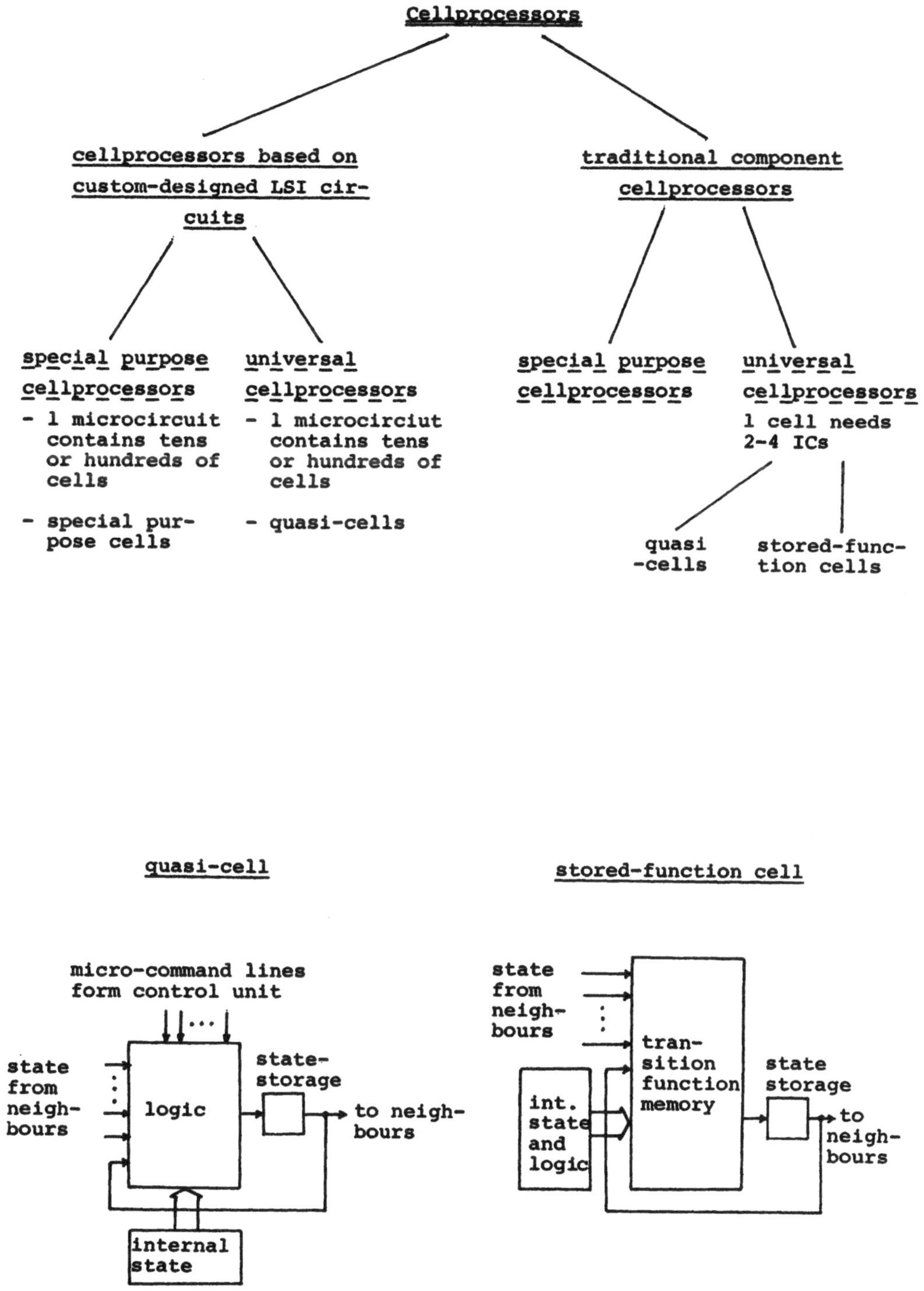

Fig.1. Classification of cellprocessors; quasi-, and stored-function cells

but high processing rates such as signal processing, control of industrial systems, picture processing, arithmetic operations and so on... In these cases cellprocessors based on standard components are effective. Now we show a traditional component based cellprocessor, and some possiblities of application.

A traditional component cellprocessor

This is a universal purpose cellprocessor, implemented from standard components /catalog-circuits/. This cellprocessor may be used as a dedicated peripheral /arithmetic processor/ of a computer system, or as a control and processing unit of an industrial system. The cellprocessor consists of three main parts i.e. cellular space, cellular space control

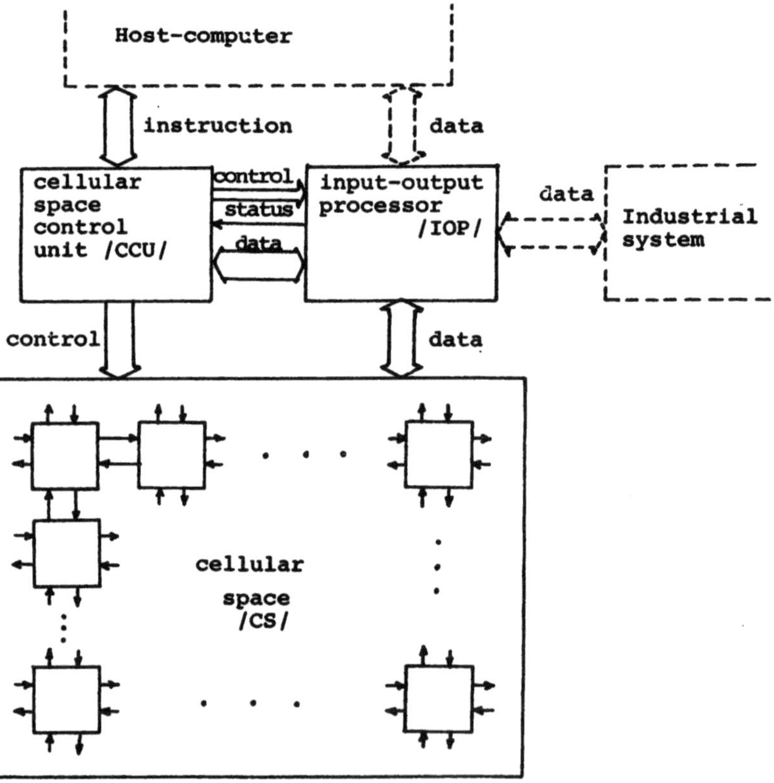

Fig. 2. Architecture of the cellprocessor

unit, and input-output processor /fig. 2/.

The two-dimensional Cellular Space /CS/ consists of 2-state stored--function cells with Neumann-neighbourhood. A cell contains a state storage, an internal-state storage, a transition function memory and some logic as shown earlier. With these cells we can emulate 4-state cells /not universally/ and NxMx2 cellular spaces /two-layer CSs/. The main property of the cellular space is that we can emulate larger CSs without overhead to get over difficulties arising from small ones. This means logically that physical cellular space rolls through the large virtual space and emulates its functions /proposed by T. Legendi/. With this capability we can solve tasks requiring larger cellular spaces-however this costs more time-, or we can build a cellprocessor with minimal hardware for a given application and I/O-rate may be tailored. The CS can work at a rate of 10 M inhomogenous cellular steps/second. The size of the cellular space depends on the requirements and modularly extensible.

The Cellular space Control Unit /CCU/ controls the cellular space and connects the cellprocessor to the host computer /interprets instructions given by the host computer and executes them by means of other parts of the cellprocessor/. The CCU is a traditional microprogrammed control unit with dynamic microprogramming.

While the CCU controls the cellular space, the Input-Output Processor /IOP/ moves data from host computer /or industrial system/ to the cellular space and vice-versa. It is a microprogrammed control unit having a buffer memory.

The parallel working CCU and IOP ensure that the cellular space can work at maximum speed.

The cellprocessor performs computations independently of host computer, exempting it from the execution of these tasks. At the beginning of the processing, the host computer loads cellprograms /cell microprograms/, and transition functions into the cellprocessor, then it is ready to perform computations. When the host computer interprets an instruction to be executed by the cellprocessor, it gives over this instruction and its parameters to the cellprocessor, and the host computer continues processing. Such an instruction points out a microroutine in the CCU, which controls the cellular space to execute the given instruction, or the cellprocessor takes over the control of an industrial system. Data transmission is continuous between host /or industrial system/, and cellprocessor. When the cellprocessor has executed the instruction of the host computer, it signals to the host.

Cellprocessor and host computer work in parallel but this parallelism and this execution method of instructions are hidden from the user.

Applications

Here we mention some application fields, where the above cellprocessor may prove effective, and fast enough. Processing rates refer to a cellprocessor containing 240 cells and requiring about 800 packages.

1; Picture processing tasks /convolution, reflection, enlargement, histogram-computation.../ can be solved about 100 times faster than with a microcomputer based on 8-bit CPU-type microprocessor.

2; Control of industrial processes /implementation of regulation algorithms/, and sequential control /evaluation of logic functions/. The cellprocessor can execute 1200 Million logic operations per second PID-regulation can be executed about 100-200 times faster than with a microcomputer based on 8-bit CPU-type microprocessor.

3; Signal processing /digital filtering.../ can be executed about 100-200 times faster than with a microcomputer based on 8-bit CPU-type microprocessor.

4; Execution of arithmetic operations /such as vector-operations/. This cellprocessor may be used as an arithmetic processor /dedicated peripheral/ of a micro-, or minicomputer.

Literature

[1] T. Legendi: Cellular systems. In this proceedings.

[2] T. Legendi: Cellprocessors in computer architecture. Computational Linguistics and Computer Languages 11. /1976/, 147-167.

[3] J. Tóth: Sejtprocesszorok tervezési kérdései. Információ Elektronika, Hungary 18, /1983/, 201-207.

[4] C. Hammer, V. Henkel, W.O. Höllelerer: BVCP-The Brunswick virtual cellular processor. Article in: Cellprocessors and Cellgorithms. Technische Universität Braunschweig, 1981.

SESSION: THEORETICAL PROBLEMS

A FEW RESULTS IN THE THEORY OF HOMOGENEOUS STRUCTURES

VICTOR ALADYEV [1)]

1. INTRODUCTION

During the recent years there has been considerable interest in the theory of homogeneous structures (HS) and their applications about which many interesting results have been obtained. Much of this work has been motivated by the growing interest in computer sciences and biological modelling. The HS (synonyms: cellular automata, iterative arrays and so on) is a formalization of the concept of an infinite regular array of identical finite-state machines uniformly interconnected in the sense that each machine can directly receive information by means of interconnecting wires from a finite number of neighbouring machines where the spatial arrangement of these neighbouring machines in the same relative to each machine in the array. Each machine can synchronously change its state at discrete time steps as a function of the states of the neighbouring machines. This function can change from time step to time step but will be identical for each machine in the array at any given time step.

The simultaneous action of these local functions will define global functions which will act on the entire array changing condigurations of machine states in the array to other configurations.

Such models have been applied in such diverse areas as pattern recognition, machine self-reproduction, cellular differentiation, evolution and development theories, theory of morphogenesis, adaptive systems, parallel algorithms and parallel processing and so on [1-8]. HS can serve as the basis for modelling of many discrete processes and they present enough interesting independent objects of investigation as well.

The modern point of view to the theory of HS has been formed under the influence of works of H. Yamada, S. Amoroso, A. R. Smith, A. Burks, H. Nishio, R. Vollmar, T. Legendi, G. Wolf and others.

In this paper we present a few new theoretical results and techniques in the theory of HS which allow to clarify such problems as nonconstructibility in HS, decomposability of global transition functions in HS and the complexity of configurations and global functions in HS. In

---

1) Tallinn-35, Paldiski mnt 171-26, P/A "Silikaat"

as much as complete proofs of this results are very long, we present a brief sketches, only. All complete proofs can be found in our latest work [4].

2. GENERAL DEFINITIONS AND NOTIONS

In the first instance, we introduce the classical concept of HS and summarize the basic definitions and notions which are used later.

The classical definition of a HS makes it an ordered set of four components

$$HS = \langle Z^d, A, \tau^{(n)}, X \rangle,$$

where A is a finite, nonempty set called the <u>state alphabet,</u> and is the set of states which each of the individual finite-state machines in the structure can assume. The state alphabet contains a distinguished element called the <u>quiescent</u> state, which is denoted by 0. Let the state alphabet $A=\{0,1,2,\ldots,a-1\}$ contain a elements. Z^d is the set of all d-tuples of integers which is used to name the <u>cells,</u> where Z is the set of integers, and is called the <u>array.</u> Each cell z in Z^d can be thought of as the name or address of the particular finite-state machine which occupies that position in the array.

X, called the neighborhood index of the HS, is an n-tuple of distinct d-tuples of integers and is used to define the neighbors of any cell, i.e., those cells from which the cell will directly receive information. The n neighbors of cell z are cell $z+\alpha_1, \ldots, z+\alpha_n$, where $X=(\alpha_1, \ldots, \alpha_n)$.

The first three above-mentioned components of any HS, namely, the state alphabet A, the array Z^d and the neighbourhood index X, form a <u>homogeneous space.</u> In order to study the operation of HS, it is necessary to have a means of describing the current state of the entire homogeneous space at any given time. The state of the entire space is called a <u>configuration</u> of the space and is simply the complete set of current states of each of the individual machines. A configuration (CF) is any mapping CF : $Z^d \to A$; null-CF($\bar{0}$) is a mapping $\bar{0} : Z^d \to 0$. C_A denotes the set of all configurations with respect to Z^d and A, i.e., $C_A = \{CF \mid CF : Z^d \to A\}$. Let c(z) be the current state of the machine located at cell z.

The support of a CF c (denoted by $[c]$) is the set of all cells z such that $c(z)\neq 0$; i.e., the support define the nonquiescent part of CF c. Configurations with finite support are of considerable interest; the set of all such configurations is denoted by $\bar{C}_A$. The set of all infinite configurations (with infinite support) is denoted by $\overset{\infty}{C}_A$, accordingly. It is obvious, $C_A=\bar{C}_A \cup \overset{\infty}{C}_A$ and $\bar{C}_A \cap \overset{\infty}{C}_A=\emptyset$ (empty set).

The operation of the HS is specified by a <u>local transition</u> function $\sigma^{(n)}$ which produces the next state of an individual machine z in terms of the states of the machines which are directly connected to z. In this work we shall be concerned with a local functions $\sigma^{(n)}$, which are defined to be a mappings from A^n to A such that $\sigma^{(n)}(0^n)=0$. For the rest, a local function is any mapping $\sigma^{(n)} : A^n \rightarrow A$. If a HS has state alphabet A and neighbourhood index $X=\{\alpha_1, \alpha_2, \ldots, \alpha_n\}$ and z is currently ib CF c, then the next state of z is determined by applying $\sigma^{(n)}$ to the set of current states of the machines in the neighbourhood of z.

The simultaneous application of a local transition function $\sigma^{(n)}$ to the neighbourhood of every cell of the homogeneous space defines a <u>global transition function</u> $\tau^{(n)}$ of the current CF c into the next CF $c\,\tau^{(n)}$.

In such a way, we introduced the classical concept of monogenic d-dimensional HS(d-HS). The operation of a monogenic d-HS($d \geq 1$) is particularly simple. If $c=c_0$ is an initial CF of the homogeneous space at time t=0, then the CF at time t=m is $c_0\,\tau^{(n)m}$, the result of applying function $^{(n)}$ to the homogeneous space m times. Let $\langle c_0 \rangle \tau(n)$ denote the configuration sequence generated by $\tau^{(n)}$ from the CF c_0, i.e., $\langle c_0 \rangle \tau(n) = \{c_0, c_0\tau^{(n)}, \ldots, c_0\tau^{(n)m}, \ldots\}$. It is said that configuration c' is predecessor of CF c if $c'\tau^{(n)}=c$.

The problem of nonconstructibility is fundamental in the study of the theoretical properties of d-HS($d \geq 1$). In [1] we introduced and investigated two different concepts of nonconstructible configurations (NCF and NCF-1).

<u>Definition 1.</u> Configuration $c \in \overline{C}_A$ is NCF for global function $\tau^{(n)}$ of d-HS($d \geq 1$) iff there does not exist CF $c' \in C_A$ such that CF $c'\ ^{(n)}$ contains no CF c as a subconfiguration.

<u>Definition 2.</u> Configuration $c \in \overline{C}_A$ is NCF-1 for global function $\tau^{(n)}$ of d-HS($d \geq 1$) iff there exists CF $\overset{\infty}{c} \in \overset{\infty}{C}_A$ such that $\overset{\infty}{c}\tau^{(n)}=c$ and there does not exist CF $c' \in \overline{C}_A$ such that $c'\tau^{(n)}=c$.

<u>Definition 3.</u> Two CF $c_1, c_2 \in \overline{C}_A (c_1 \neq c_2)$ form for global function $\tau^{(n)}$ of d-HS($d \geq 1$) a pair of the mutually erasable CF(MEC) iff $c_1\tau^{(n)} = c_2\tau^{(n)}$.

Now we proceed to the discussion of the general results of the paper. The detailed discussion of the general definitions, notions and concepts of d-HS($d \geq 1$) can be founded in our monograph [1].

3. GENERAL RESULTS

The complexity is one of the most intriguing and vague concepts in most cases. One diffeculty in dealing with this problem is that we need a satisfactory measure for complexity. Investigations in this direction would be extremely desirable. In [1] we introduced a concept of complexity for finite configutations in d-HS($d \geq 1$). At present we known three approaches to the definition of concept "the quantity of information" : combinatorical, probabilistic and algorithmical, which use the theory of recoursive functions. For the last case N. Kolmogorov defined the relative complexity of some object A (comparatively of object B) by the minimum length of programme of deriving of A from B. In the capacity of representatives of objects Kolmogorov chooses their binary numbers in some numeration and in the capacity of programme Turing machine programme can be choosen [1,7].

Our approaches can be also called algorithmical but its differ from Kolmogorov's one. The essence of our concept of complexity consist in the estimation of complexity of growing of arbitrary finite CF from some primitive CF o_p by means of the finite number of global functions from some set T_f.

In [1,7] a number of interesting results about some problems of complexity observed in 1-HS with any number of states are described. For example, the essential difference between Kolmogorov's concept of complexity of finite objects and our one was established. From Kolmogorov's concept may be drawn the following conclusion : the set C_B of all binary 1-dimensional finite CF consists of configurations of the limited complexity, only. Whereas our approach establish the existence in C_B of configurations of any complexity. The introduced concept of complexity present appointed interest for investigations of the dynamic properties of d-HS, including parallel languages generated by d-HS($d \geq 1$). But this concept was based on a hypothesis 3[1]. The following theorem proves this hypothesis.

<u>Theorem 1.</u> For any finite set T of global functions of d-HS($d \geq 1$) and any finite subset $G \subset \overline{C}_A$ the following correlation take place :

$$\bigcup_{\substack{c \in G \\ \tau \in T}} \langle c \rangle_{\tau} \subset \overline{C}_A$$

<u>Proof.</u> The proof of this theorem is based on our previous results in nonconstructibility in d-HS [1,7]. Suppose, that there exist some set $T=\{T_1, T_2, \ldots, T_q\}$ of global functions of d-HS($d \geq 1$) and configurations $c_i \in C_A (i=\overline{1,q})$ such that

$$\bigcup_i \langle c_i \rangle_{T_i} = \overline{C}_A \qquad (1)$$

Using our results in nonconstructibility in d-HS($d \geq 1$) [1] we proved that the set of all functions $T_i (i=\overline{1,q})$ in (1) is divided into three subsets W, L and M such that :

- W : contains functions $T_i(i=1,p)$ which has NCF;
- L : contains functions $T_j(j=p+1,l)$ which has NCF-1 without NCF;
- M : contains functions $T_k(k=l+1,q)$ which has no NCF and NCF-1, generally speaking. On the basis of sets N_i of NCF for each function

$T_i \in W$ is constructed the finite CF o_W which contains p NCF from N_i($i = \overline{1,p}$). Such CF o_W is the generalized NCF for any global function $T_i \in W$, i. e., CF o_W is NCF for each function T_i($i=\overline{1,p}$). In view of the properties of NCF in d-HS($d > 1$) [1], it can be easily seen that there exist the infinite set of configurations of type o_W for global functions T_i.

Consider now global functions T_j($j=\overline{p+1,l}$) from L. By definition, each global function $T_j \in L$ has NCF-1 without NCF. It is well known [1], that a set Q_j of all NCF-1 u_j^m for any global function T_j is infinite ($j=\overline{p+1,l}$; $m=1,2...$). Let o_L be d-dimensional ($d > 1$) CF of the following type : in the opposite points of diameters of d-dimensional sphere are placed in pairs CF u_j^m; on the sphere of large enough diameter all CF u_j^m are placed wide apart. It is easily verified that CF o_L is the generalized NCF-1 for any global function $T_j \in L$, i. e., CF o_L is NCF-1 for each function T_j($j=\overline{p+1,l}$). Furthermore, there exist the infinite set of configurations of type o_L for global functions T_j.

Disposing the CF o_W in the centre of CF o_L, we obtain new CF o_{WL}. It can be proved, that set WL of configurations of type o_{WL} is infinite and each CF from WL is NCF and/or NCF-1 for global functions T_j($j=\overline{1,l}$). Consequently, configurations from set WL cannot be generated with the aid of the finite number of global functions T_j, using arbitrary initial finite configurations o_j($j=\overline{1,l}$). Thus, we have only global functions from set M left to consider. But it is well known [1] that any global function (which has no NCF and NCF-1) from set M possesses extremely weak generation possibilities. From detailed analyses of the generation possibilities of such global functions from M follows that set WL of the finite configurations cannot be generated with the aid of the finite number of global functions T_k, using arbitrary initial finite configurations o_k ($k=\overline{l+1,q}$). Thus, we obtained contradiction with our supposition about possibility of correlation (1). This contradiction proves the theorem for case of dimensionality $d > 1$. For case of dimensionality $d=1$ the proof demands some modification. This situation is linked with nonequivalence od 1- and d-HS($d \geq 2$) with respect to a number of characteristics, including techniques of investigations [1,7] . Q. E. D.

Closely connected to this theorem is the problem of existence of the universal configurations (UCF) in HS. We presented a solution of the problem for the simplest cases [1]. Theorem 1 will make it possible to consider the problem from another angle. Furthermore, the theorem is extremely important for theoretical investigations of dynamic properties of d-HS($d \geq 1$).

Questions of nonconstructibility are fundamental problems in the study of the theoretical properties of d-HS. Since the study of histories of configurations in d-HS play general role in investigations of the dynamic properties of d-HS, it is interesting to have conditions of existence of nonconstructible configurations in d-HS($d > 1$). In [1] we introduced two different concepts of nonconstructible configurations (NCF and

NCF-1). Questions of the existence of NCF in d-HS($d \geq 1$) were studied in detail [1]. For 1-HS a number of interesting results we proved with respect to NCF-1 [1,7]. Since d-HS has four possibilities with respect to the existence in it NCF and NCF-1 (the full set of possibilities), it is desirable to obtain satisfactory criteria of these possibilities. In the case of NCF we have a few criteria [1]. Theorem 3.14[1] gives a more convenient criterion of the existence of NCF-1 for 1-dimensional global functions $\tau^{(n)}$ without NCF. This result can be extended on the arbitrary d-HS($d \geq 1$), too.

Theorem 2. d-HS($d \geq 1$) without NCF possesses NCF-1 iff for global function $\tau^{(n)}$ of d-HS there exists CF $\overset{\infty}{c} \in \overset{\infty}{C}_A$ such that $\overset{\infty}{c}\tau^{(n)}=\bar{0}$.

Proof. For case of dimensionality d=1 theorem 2 was proved in monograph [1] as theorem 3.14. Without loss of community we will consider 2-HS with Moore's neighbourhood index. Let arbitrary 2-HS without NCF has global function $\tau^{(n)}$ such that for it there exists NCF-1. On the basis of definition 2 it is not hard to prove that for $\tau^{(n)}$ there exists CF $\overset{\infty}{c} \in \overset{\infty}{C}_A$ such that $\overset{\infty}{c}\tau^{(n)}=\bar{0}$.

Let now for global function of 2-HS without NCF there exist CF $\overset{\infty}{c} \in \overset{\infty}{C}_A$ such that $\overset{\infty}{c}\tau^{(n)}=\bar{0}$. Suppose, that such 2-HS possesses no NCF-1. On the basis of the supposition it can be shown [4] that global function $\tau^{(n)}$ of such 2-HS must satisfy the following two conditions :

- each square subconfiguration of large enough size of CF $\overset{\infty}{c}$ is transformed by function $\tau^{(n)}$ into CF of type "indissoluble circle" : from such CF cannot be selected separate independent finite subconfigurations;
- each CF $c \in \bar{C}_A$ with respect to function $\tau^{(n)}$ has only predecessor from set $\bar{C}_A$ and only predecessor from set $\overset{\infty}{C}_A$.

In this connection it is necessary to note, that for case of 1-HS there exists no global functions for which take place the second condition (2).

On the other hand, it can be shown that each 2-HS, on condition that take place given supposition, break at any rate an above-mentioned condition (2). The obtained contradiction proves the theorem. Q. E. D.

Thus, our previous results in nonconstructibility in d-HS and theorem 2 present quite satisfactory criteria of the existence of NCF and NCF-1 without NCF in d-HS($d \geq 1$). Theorem 2 allows to clear the essence of nonconstructibility in d-HS with the greater detailed elaboration. However, it has a series of unsolved problems. It is desirable to receive similar criteria of the existence of NCF-1 in d-HS in the general case. The solution of this problem is absent up to this point even for case of 1-HS.

Decidability results play extremely important role in the theory of d-HS. Closely connected to theorem 2 is the following decidability problem : Is it decidable whether an arbitrary d-dimensional global function $\tau^{(n)}$ has CF $\overset{\infty}{c} \in \overset{\infty}{C}_A$ such that $\overset{\infty}{c}\tau^{(n)}=\bar{0}$? The solution of the problem for case of dimensionality d=1 was given in work [1]. For case of dimensionality $d \geqslant 1$ we have the following result.

Theorem 3. The problem of the existence for d-dimensional global function $\tau^{(n)}$ of CF $\overset{\infty}{c} \in \overset{\infty}{C}_A$ such that $\overset{\infty}{c}\,\tau^{(n)} = \overline{0}$ is algorithmically solvable.

Proof. For case of dimensionality d=1 the theorem was proved in our work [7] as theorem 5. Let $\tau^{(n)}$ be an arbitrary global function of dimensionality $d > 1$ in alphabet A. On the basis of the correlations $\sigma^{(n)}(x_1, x_2, \ldots, x_n) = 0$ $(x_i \in A;\ i = \overline{1,n})$ for local function $\sigma^{(n)}$ defined by global function $\tau^{(n)}$ all configurations $c_1 \in C_A$ such that $c_1 \tau^{(n)} = \overline{0}$ is constructed. It can be shown that such procedure of growing of each predecessor for CF $\overline{0}$ demands no more than D steps, where value D depends on the neighbourhood index X and alphabet A of d-HS($d > 1$). After D steps it is decidable whether or not growing CF $c_1 \in \overset{\infty}{C}_A$. Q. E. D.

A number of authors have deal with the well-known General Decomposition Problem (GDP). The GDP can be formulated as follows : Can any global function $\tau^{(n)}$ of d-HS($d \geqslant 1$) be presented in the form of composition of the finite number of more simple global functions $\tau^{(n_i)}$ in the same alphabet, i. e.,

$$\tau^{(n)} = \tau^{(n_1)}\,\tau^{(n_2)} \ldots\ldots\, \tau^{(n_k)} \qquad (n_i < n;\ i = \overline{1,k}) \qquad (3)$$

The GDP in d-HS($d \geqslant 1$) was studied in detail enough [1,3,7], using our results in nonconstructibility in d-HS [1], Yamada-Amoroso's and Yamada-Kaoru's results in completeness problem, with the help of Shannon's function, and using results in K-valued($K \geqslant 2$) logics [7]. In this paper the problem receives further decision on the basis of other approaches.

Theorem 4. There exists the infinite set of global functions $\tau^{(n)}$ of binary 1-HS which have not representation (3).

Proof. Assume that for any integer $n \geqslant 3$ each binary function $\tau^{(n)}$ has representation (3). Then, with the aid of descending reduction the following representation for function $\tau^{(n)}$ can be easily derived from (3) :

$$\tau^{(n)} = \tau^{(2)}_{i_1}\ \tau^{(2)}_{i_2}\ \ldots\ldots\ \tau^{(2)}_{i_m}\ ;\quad \tau^{(2)}_{i_j} \in M^2 \qquad (4)$$
$$(j = \overline{1,m}),$$

where M^2 is the set of all 1-dimensional binary global functions $\tau^{(2)}$ with neighbourhood index $X = \{0, 1\}$. It is obvious, the set M^2 contains only eight global functions. From the detailed analysis of these functions follows that four functions have NCF, a function has NCF-1, and three functions are meaningless for representation (4). Using these results [4], we can prove impossibility of the representation (4) for infinite set of global functions $\tau^{(n)}$, to say the least by the third methods. These contradictions allow to prove the theorem. Q. E. D.

I hope that this work will help to clear up some aspects to the mathematical theory of d-HS($d \geqslant 1$) as well as giving some information about our latest investigations in this direction.

Literature

1 Aladyev, V.: Mathematical Theory of Homogeneous Structures and Their Applications. Valgus Press, Tallinn 1980.

2 Aladyev, V. et al: Mathematical Biology of Development. Nauka, Moscow 1982.

3 Aladyev, V.: New results in the theory of homogeneous structures. Beiträge zur Theorie der Polyautomaten, Informatik-Skripten.8, Braunschweig (1984), 3-15.

4 Aladyev, V.: The Recent Investigations in the Theory of Homogeneous Structures. P/A "Silikaat", Tallinn (submitted for publication).

5 Cellprocessors and Cellalgorithms (Ed. by R. Vollmar and T. Legendi), Common Research Report, no. 1, Braunschweig 1981.

6 Legendi, T.: Cellprocessors in Computer Architecture. Comp. Linguistics and Comp. Languages XI (1976), 147-167.

7 Parallel Processing Systems (Ed. by V. Aladyev). Valgus Press, Tallinn 1983.

8 Vollmar, R.: Algorithmen in Zellularautomaten. B.G. Teubner, Stuttgart 1979.

PARALLEL PROCESSING STRUCTURES FOR THE INVARIANT REPRESENTATION AND RECOGNITION OF PATTERNS

Ireneusz H. Defée [1]

1. Introduction

Representation of visual space is of primary importance for the construction of intelligent systems and robots. This problem is not solved in satisfactory manner also extensive work has been done in the past on analytical and structural methods of invariant representation [1], [2], [3], [4]. Recently, a progress has been in the area of low-level computer vision dealing with stereopsis and shape from shading [5], [6], [7]. Algorithms presented are robust, based on local parallel computations, but not universal - using some "reasonable" assumptions about visual world.

In this paper, we try to derive the structure of a system capable of building of a invariant representation of objects placed in visual space using similar approach. We assume that the system operates exclusively with local parallel computations and its task is local representation of each object of visual space invariant w.r.t. translations, rotations and scale changes. The solution is obtained by considering the action of these transformations on various classes of patterns. In result, generalized pyramidal structure is derived, performing low pass filtering, detection of local invariants and synthesizing invariant representation. Representation of complex patterns is realized by modulating the density of input receptors. In this way, detailed representation is built only for limited area of patterns, and at the same time general representation is obtained. Using parallel/serial operation, detailed representation for patterns with practically unlimited complexity can be synthesized.

2. Representation and invariance

Representation is a formal procedure making explicit some properties or information contained in the system on which it operates. The system considered here is a visual space, it will not be formalized but treated intuitively as composed of certain configurations of objects and transformations acting on them. Each configuration of objects is called scene. It is assumed that scenes are projected onto a finite grid of receptors and their representations synthesized by a cellular system using only local computations. Because of the locality condition, the goal of building of a representation is obtaining local information about every aspect of visual scene. A primary objective here is that the representation should be invariant w.r.t. the variety of transformations and

[1] Department of Electrical Engineering, Technical University of Szczecin, 70-313 Szczecin, Av. Sikorski 37, Poland

computationally effective.

We consider first simplified model dealing with two-dimensional scenes projected onto a finite grid of receptors and translations (T), rotations (R) and scale changes (S) acting as transformations of objects. Two classes of objects are treated separately: local, with size being within one or few steps of parallel computations and nonlocal, or global, extending over some orders of magnitude of parallel computations. It is necessary also to emphasize some geometrical properties of transformations acting on objects:

- Translations - all points treated uniformly and all geometrical properties preserved,
- Rotations - points are moved by a constant angle, but not a constant distance; the center is a fixed point, usually not known in advance. All geometrical properties are preserved.
- Scale changes - distances between points of objects are changed by a constant factor, thus only angles and curvatures are local invariants. Because of the finitness of receptor grid, amount of recoverable information is not constant.

It is seen from this description that scale change is fundamentally different transformation from the other two. We thus consider first local patterns with translations and rotations acting on them. Simplest local pattern is single point O, which may appear anywhere in the visual space. The task of a system is localization of the point (stereopsis) and invariant representation, which means in this case generating a signal regardless any translation T(O). From the condition of locality, such a representation is realized by the pyramid composed of detectors and tree network shown at Fig. 1.

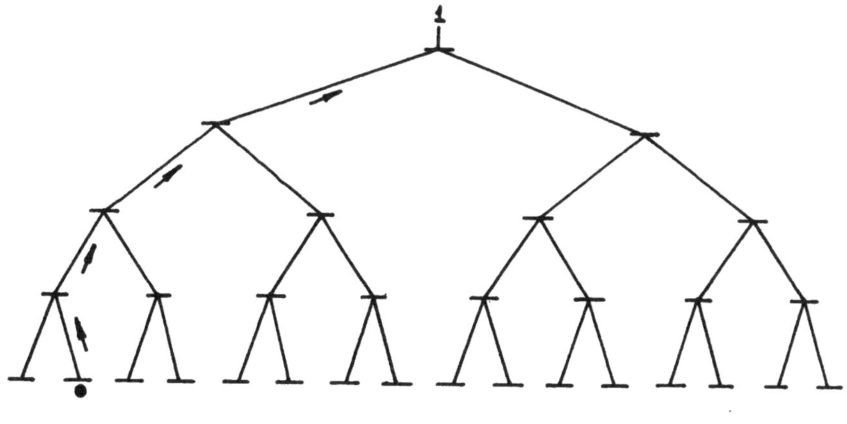

Fig.1. Tree network for local translational invariance

Assume now that by an application of smallest resolvable scale increasing transformation, the point O has evolved into some local form S(O). Because of finite resolution and local computations, independent detectors must exist for each local form S(O). By an application of translation T(S(O)) it is seen that invariant representation is realized by tree networks operating on detectors of local forms and analogous to the one shown at Fig.1. To ensure rotational invariance, one may note that for local forms it is sufficient to consider only rotations about local center because of small size of local forms. Rotational invariance is realized then by identifying outputs from trees for rotated invariants.

From this analysis, the parallel processing structure for local forms is composed from layer of single point detectors and layers of detectors of local forms at the bottom level. Next levels are realized by separate interleaved tree networks, composed of OR-like elements and at the highest level, signals from rotated forms are identified by another set of such elements. We can see thus, that for local objects, the problem of invariant representation is solved easily. However, nonlocal forms are much more difficult because of their practically unlimited variety and complexity. To resolve this, we limit ourselves to highly structured patterns which are defined later, and consider the effect of scale operator on objects. First effect, which must be considered when scale is increased to change from local to nonlocal form, is sampling. Increasing the scale increases the physical size of objects and the number of samples taken by a receptor grid. From the signal theory point of view, this is described by increased spatial frequency bandwidth. However, because physical size is changed, there is also a relative shift of a frequency bandwidth on frequency axis. Thus, when a signal has a spatial frequency band ΔF and its scale is increased by a factor s, then the frequency band is shifted down by $\Delta F/s$ and new frequencies become recoverable increasing the band to $\Delta F \cdot (1 + 1/s)$. Consider now how the invariance can be preserved for such a kind of scale changes. Since the effect is described by shifting the frequency band, invariance is realized by a bank of constant-Q bandpass filters with identical characteristics. Input signal is decomposed according to its frequency content and filter outputs are measured. When there is a change of scale it results only in a shift of filter outputs in a bank. In two dimensions, the condition of local processing means that the filters must be connected in a cascade form. Closer description of such a cascade is obtained by specifying closer filter requirements. Natural such requirements are: a) constant-Q bank which means uniform characteristic, b) convolutional closedness, which means preservation of filtering characteristic by a cascade, c) optimal position/spatial frequency resolution and d) simple realization by cascaded local computations in the position domain.

Conditions c) and b) imply here the cascade of bandpass Laplacian filters. Moreover, it is known also [8] that bandpass Laplacians are well aproximated by differences of low-pass Gaussian filters with different cut-off frequencies which have also unique property which allows for their realization by Gaussian weighted local averaging in the position domain. Such a cascade is shown at Fig.2. Differences between subsequent levels make bandpass filters. Outputs from bandpass filters at different levels are transmitted by a tree network. Next, filter outputs from different levels are compared to detect shift in a frequency band.

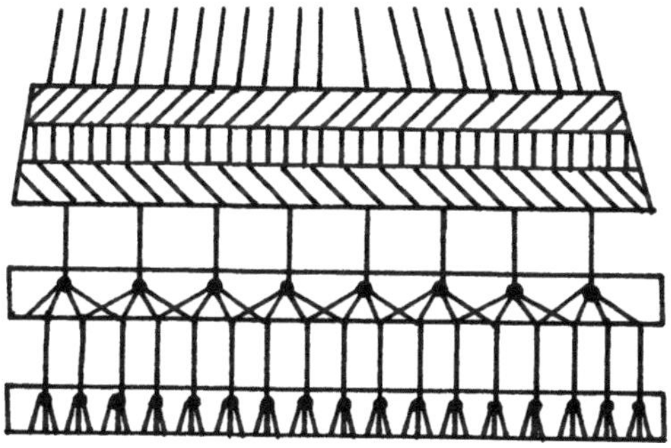

Fig.2. Gaussian filtering cascade

3. Global processing

Until now we considered simplest cases of invariant representation. The problem is approached for complex, nonlocal patterns by considering invariants of scale changes. Since, as mentioned earlier, such invariants are made locally by curvatures and angles, first task is the computation of oriented edges. A useful hint here is that optimal filters for edge detection are well approximated by bandpass Laplacians [9], which allows for use of the cascade described earlier. Since orientation of edges is necessary here, we note that orientation can be obtained directly in a local parallel processing by the use of filters based on polar (r,φ) or logarithmic-polar $(\ln r,\varphi)$ coordinate system. Using orientations, local invariants are obtained by the computation of angles and local curvatures. Note here, that invariant representation may be realized for angles and curvatures in the same way as for local forms.
Now we shall proceed to the computation of an invariant representation for global patterns using invariants. Consider full range of action of scale decreasing transformation operating on invariants. At its certain value, every two invariants become local and can be merged. From the

point of pattern description, important are those invariants which have common orientation line or line of equal curvature. We need thus computations along such lines for merging of invariants. This is done by local parallel computations using tree structures similar to the shown at Fig.1. Main step in the construction of an invariant representation is as follows. Assume that two invariants were merged into local form and that special merging detectors exist.
The computations proceed now along lines of equal orientation/curvature with signals from merging detectors. In this way, signals from merging detectors having a common orientation line will be merged and detected. This procedure can be repeated, allowing thus for synthesizing local representation for any number of invariants provided that the processing network is sufficiently built up. One may note here that merging detectors operate on exponentially increasing number of invariants. This, however, does not resolve fully the problem of complexity of objects and their representation.

4. Representation of complex objects

The structure described so far performs computation of local forms and invariants on various scale levels, and builts representation by merging invariants into local forms using tree networks such as shown at Fig.1. Operations performed by this kind of generalized pyramidal structure are similar to scale reduction realized on various levels of complexity: points, local forms,spatial frequencies, invariants. Because the number of processing elements is always decreasing exponentially and all operations are local, the structure is structurally and computationally effective. But it is not universal since in fact the processing is based on implicit assumption that objects preserve their invariance properties within certain range of scale reductions and projections onto a finite grid of receptors. Objects of such a kind may be called structured as opposed to unstructured ones, for example randomly distributed dots.

Another problem is made by complexity of objects having large number of local invariants. Representation of such objects would require large number of detectors for merged invariants, outside of reasonable limits. This problem can be solved by the modulation of receptor density across the input array (Fig.3). The density of receptors is high and sufficient for building detailed representation only in the small area in the middle of receptor array. Outside of this area it falls exponentially which results in linear decrease of resolution. Such a modification allows for the synthesis o general invariant representation of structured objects using big, low-resolution area. At the same time, detailed representation may be realized by serial/parallel operation using middle of the array. In this way, complexity of a system is greatly

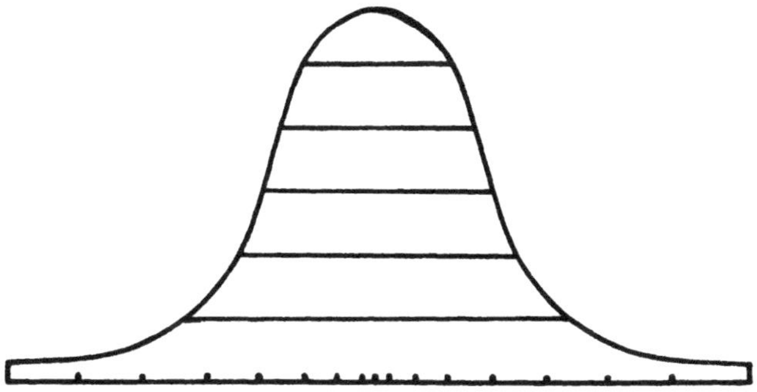

Fig.3. Exponential modulation of receptor density

reduced, but objects with practically unlimited complexity may be processed.

5. Conclusion

The parallel processing structure for the invariant representation and recognition has been developed by considering various classes of patterns and transformations acting upon them. The structure has a form of generalized pyramid which is both computationally and structurally effective. It is not universal but amenable rather for structured objects with limited complexity of parallel processing, which is partially compensated by the possibility of serial/parallel operation. Further research is directed into the extension of invariance to broader class of topological transformations and applications to the synthesis of intelligent systems and robots.

6. References

[1] T.Pavlidis, "A review of algorithms for shape analysis", Comp. Graphics Image Processing, Vol.7, pp.243-258, 1978.

[2] T.Pavlidis, "Algorithms for shape analysis of contours and waveforms", IEEE Trans.Pattern Anal.Machine Intell. Vol. PAMI-1, pp.301-311, 1980.

[3] M.G.Thomason, "Syntactic-semantic techniques in pattern recongition: a survey", Intern.J.Comp. Inform.Sci., Vol.11, pp.78-100, 1982.

[4] L.Jacobson and H.Wechster, "A theory for invariant object recognition in the frontoparallel plane", IEEE Trans. Pattern Anal.Machine Intell., Vol.PAMI-6, pp.325-331, 1984.

[5] D.H.Ballard et.al, "Parallel visual computation", Nature, Vol.306, 5938, pp.21-26, 1983.

[6] D.Marr and T.Poggio, "A computational theory of human stereo vision", Proc. Royal Soc.London B, Vol.204, pp.301-328, 1979.
[7] A.P.Pentland, "Local shading analysis", IEEE Trans.Pattern Anal. Machine Intell. Vol. PAMI-6, pp.170-187, 1984.
[8] P.J.Burt, "Fast filters transforms for image processing", Comp. Graphics Image Processing, Vol. 16, pp.20-51, 1981.
[9] J.D.Daugman, "Six formal properties of two-dimensional anisotropic visual filters", IEEE Trans. Syst. Man, Cybern., Vol. SMC-13, pp.882-887, 1983.

PARSIS - A PARALLEL SIMULATION SYSTEM

Reinhard Klette[1)] and Roland Heinrich [2)]

Abstract. PARSIS represents a general framework for simulating SIMD techniques where different interconnection patterns for the set of processing elements may be fixed. In this paper, some topics in the theoratical work connected to PARSIS will be dealt with, regarding the description of the used processor masking scheme, the analysis of lower time bounds by data transfer descriptions, and the design of image processing algorithms for a specific class of pictures (cellular structure representation by binary pictures). Concrete results will be given for the sqare, orthogonal interconnection network

0. INTRODUCTION

During the last years some parallel simulation systems have uprisen supporting efficient design of parallel software and hardware, e.g. [6]. The general structure of PARSIS was outlined in [4], the first phase of the implementation is reported in [3]. A short note about this project level was published with [5]. As explained in [3,4,5], the flexible concept of PARSIS allows the implementation of different interconnection functions defining the neighborhood relations between processing elements (PEs). According to the main orientation of this Workshop, emphasis is given in this contribution to the SQUARE network that is closely related to cellular automata based on the same interconnection structure [8] but different in processing power for the single processing element. Also, in our system each PE is informed about its position within the processor array by knowing its "name", i.e. its index or address within the array, and may be active or idle depending upon its index matches a given address or not, respectively.

The following topics will be considered in this paper: In Section 1, the general structure of PARSIS will be explained briefly. In Section 2, the patterns of activated PEs on the SQUARE network produced by address masks will be described. In Section 3, the general method to analyze lower time bounds by quantitative measures of data transfer will be explained and illustrated. Finally, in Section 4 the time complexity of some parallel algorithms on the SQUARE system will be compared with lower time bounds by data transfer analysis.

---

1) Academy of Sciences, Central Institute of Cybernetics and Information Processes, Department of Artificial Intelligence, Kurstr. 33, DDR-1086 Berlin, German Democratic Republic

2) Friedrich-Schiller-University, Department of Mathematics, University Tower 17th Floor, DDR-6900 Jena, German Democratic Republic

1. PARSIS & PARSIL

The general SIMD model as implemented by PARSIS is characterized by a finite set of PEs, an interconnection network, and a central processing unit (CPU). The CPU has a (central) random access memory, $R_{CPU}(0)$, $R_{CPU}(1)$, ... , with a distinguished accumulator $R_{CPU}(0)$. We shall use $R_{CPU}(j)$ for denoting the register or the contents of this register, as well, depending upon the context. The CPU spreads a single instruction stream to the synchronised working PEs having (local) random access memories, R(o,0), R(i,1), ... , with a distinguished accumulator R(i,0), for processor with index or address i. Each PE knows its index. Let ind be the set of all PE indeces. Each PE is capable of performing some basic operations which take place in its accumulator. Direct data access is restricted to its own registers, to the accumulators of the directly connected PEs in the sense of the given interconnection network, and, possibly, to the accumulator of the CPU. Any data transfer between CPU and PEs is restricted to serial mode in PARSIS, but there may be parallel I/O channels to the array of PEs. An interconnection function f maps ind into ind $\cup \{D\}$, where D denotes the dummy address. The network is characterized by p different interconnection functions $f_o, f_1, \dots, f_{p-1}$ on ind where PEs j and k are directly connected iff there exists an i, $0 \le i \le p-1$, such that $f_i(j)=k$, cp. [7]. The SQUARE network for an array of N x N PEs is characterized by four interconnection functions n,w,s,e with

$$n(i) = \begin{cases} i-N \text{ , if } N \le i \\ D \text{ otherwise,} \end{cases} \qquad w(i) = \begin{cases} i-1 \text{ , if } i \ne 0 \text{(mod)} N \\ D \text{ otherwise,} \end{cases}$$

$$s(i) = \begin{cases} i+N \text{ , if } i < N^2-N \\ D \text{ otherwise, and} \end{cases} \qquad e(i) = \begin{cases} i+1 \text{ , if } i \ne N-1 \text{(mod)} N \\ D \text{ otherwise,} \end{cases}$$

for $i = 0,1,2,\dots,N^2-1$. Collections of examples of networks are contained in [4,7].

The language PARSIL for simulating parallel processes, basically consits of serial and parallel instructions. Part of any parallel instruction addressed to the PEs is an enable/disable mask to select a subset of PEs that are to perform the given instruction, the remaining PEs will be idle. As standard masks simple bit patterns $[A_n \dots A_1 A_o]$ for PE indeces are used, $A_k \in \{\overline{0}, \overline{1}, X\}$, $\overline{0} = \{0\}$, $\overline{1} = \{1\}$, $X = \{0,1\}$ for $k = 0,1,\dots,n$. Index i matches a bit pattern $[A_n \dots A_1 A_o]$, i.e. PE with index i will be activated by this mask, iff there exist values $a_k \in A_k$ for $k = 0,1,\dots,n$ such that $i = a_o 2^o + a_1 2^1 + \dots + a_n 2^n$. For simplification, leading X's may be omited in these masks, i.e. masks $[A_m \dots \dots A_1 A_o]$ and $[X \dots XXA_m \dots A_1 A_o]$ are identified. In the extreme case, by $[X \dots XX] = [X] = []$ all PEs are activated. For example, all activated (•) and idle (o) PEs of a 16 x 16 SQUARE system are in the following positions if masks [1] or [OXXXXXXX] are used, respectively:

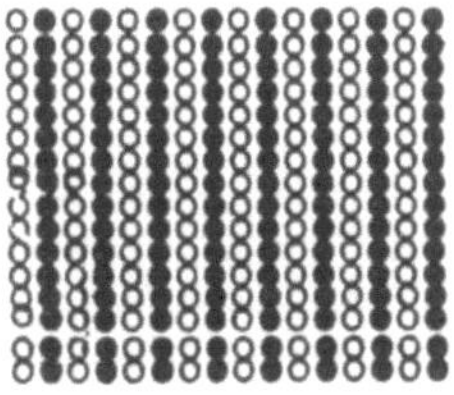

[1]

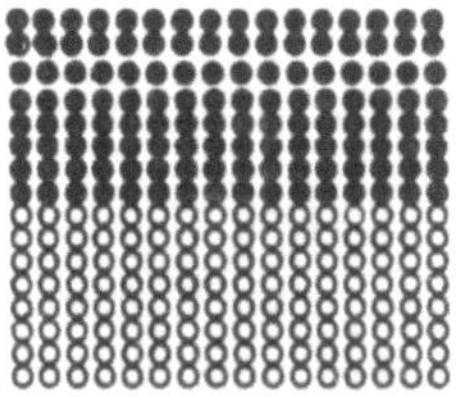

[0xxxxxxx]

Any parallel instruction consists of a PE address mask, an operation code, and an operation address. In case of the SQUARE system working in off-line mode (the PE accumulators are fixed as input+output registers of the array, during the solution of one problem no input or output is assumed) the following operations were implemented in PARSIS: STORE, ADD, SUB and MULT with addresses m or $:i_1,\dots,i_l$, and LOAD, DIV, and SGN with addresses m or :i. Here, address m denotes register R(j,m), $:i_1,\dots,i_l$ accumulators $R(f_{i_1}(j),0),\dots,R(f_{i_l}(j),0)$, and :i accumulator $R(f_i(j),0)$, for interconnection functions $f_0,\dots,f_{p-1}$ and any activated PE with index j. For example, by

mask ADD m

the contents of register m is added to the contents of the accumulator for all PEs activated by mask. By SGN the function sign(x) was implemented, with sign(x) = 1 if $x \geqslant 0$, and = 0 otherwise. For example, max(a,b) = b + (a-b)sign(a-b) and min(a,b) = a - (a-b)sign(a-b). Any serial instruction concists of an operation code and an operation address, for SQUARE the operations LOAD (with addresses m, =x, and (j)), STORE (with addresses m and (j)), and ADD, SUB (with address m) were implemented. Here, m denotes register $R_{CPU}(m)$, =x constant x, and (j) accumulator R(j,0). Finally, by instructions JGTZ label, loops are possible in PARSIL programs for the SQUARE system.

<u>Example 1.</u> For illustration of PARSIL, let us consider the following simple computational problem. At the beginning, a N x N binary image f, $f(x,y) \in \{0,1\}$ is located in the array, R(xN+y,0) = f(x,y). By the parallel program, all isolated points with value 1 have to be replaced by points with value 0, where "isolated" is assumed with respect to the 8-neighborhood {(1,0),(-1,0),(0,1),(0,-1),(1,1),(-1,-1),(1,-1,(-1, 1)}. A first draft of the parallel algorithm might be

<u>begin</u>

R(i,1) := R(o,0) for all PEs,

R(i,0) := R(i,0) + sum of accumulators in 8-neighborhood,

<u>if</u> R(i,0) = 1 <u>then</u> R(i,0):= 0 <u>else</u> R(i,0) := 1 <u>endif</u>

R(i,0) := R(i,0) x R(i,1)

<u>end</u>

In this algorithm a parallel <u>if</u>-<u>then</u>-<u>else</u>-construction is used which is not contained in PARSIL as basic instruction. But, the question "$R(i,0) \leq c$?" may be answered by the following sequence of PARSIL instructions:

[] SUB 2 (value c+1 in R(i,2) at this moment)
[] SGN 0
[] STORE 1

If R(i,0) ≤ c, then R(i,1) = 0 , if R(i,0) > c , then R(i,1) = 1 at this moment. Finally, by the sequence

[] SUB 0
[] SUB 2
[] SGN 0
[] STORE 1

it follows that R(i,1) =1 iff R(i,0) ≤ c at the beginning, and R(i,1) = 0 otherwise.

Example 2. Assume that at a certain moment the maximum has to be computed within the 4-neighborhood {(1,0),(-1,0),(0,1),(0,-1)} for the array of PE accumulators in parallel. This can be done by using a subroutine for max (using sign, see above), where all PEs are performing the following sequence of parallel instructions: R(i,1) := R(i,0), R(i,2) := R(s(i),0), R(i,3) := R(e(i),0), R(i,4) := R(n(i),0), R(i,5) := R(w(i),0), R(i,0) := max(R(i,1),R(i,2)), R(i,0) := max(R(i,0),R(i,3)), R(i,0) := max(R(i,0),R(i,4)), R(i,0) := max(R(i,0),R(i,5)). Altogether, 33 basic operations in PARSIS are sufficient for performing this parallel computation of maximal values within 4-neighborhoods.

In the sequel, the number of basic PARSIS operations is used as a measure for describing the time of a program. Of course, this is quite a rough measure because all instructions are assumed to have equal performance time 1.

2. MASKS ON SQUARE

For writing programs in PARSIL the possibilities of the masking scheme can be used efficiently if the structure of patterns is transparent which are produced on the array by concrete masks dividing the PEs in activated and idle ones. For example, for performing geometric transforms within the array the masking scheme can be applied for abtaining fast parallel solutions. In such moments, it is necessary to decide wheter patterns as

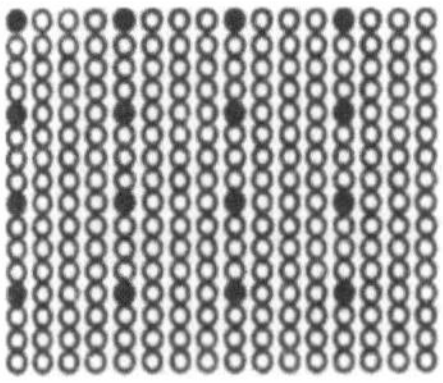

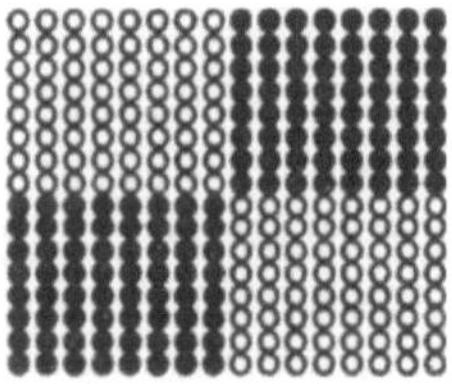

are possible by this masking scheme, and if "yes", what mask is producing a given pattern. For that reason, in [3] the masking scheme on SQUARE was analyzed with algebraic and algorithmic means.

Let $\mathfrak{M}_n$ be the set of all binary matrices $(p_{ij})_{i,j=0,1,\ldots,n-1}$ with $p_{ij}\in\{0,1\}$. For $n = 2^m$ let $M = (p_{ij})_{i,j=0,1,\ldots,n-1}$ be an element of $\mathfrak{M}_n$, let $1\le a_1\le m$ and $1\le a_2,a_3\le 2^{a_1}$. Then, submatrices of M are defined by $M_{a_1 a_2 a_3} := (p_{ij})_{i\in I, j\in J}$, where

$I = \{i:\ (a_2 - 1)2^{m-a_1} \le i \le a_2 2^{m-a_1} - 1\}$, and
$J = \{j:\ (a_3 - 1)2^{m-a_1} \le j \le a_3 2^{m-a_1} - 1\}$.

For example, if $m\geqslant 2$ and $a_1 = 2$, then the following partition results

$$\begin{matrix} M_{211} & M_{212} & M_{213} & M_{214} \\ M_{221} & M_{222} & M_{223} & M_{224} \\ M_{231} & M_{232} & M_{233} & M_{234} \\ M_{241} & M_{242} & M_{243} & M_{244} \end{matrix} .$$

Furthermore, for $M\in\mathfrak{M}_n$, $n = 2^m$, $I_n := \{(a_1,a_2,a_3):\ 1\le a_1\le m\ \&\ 1\le a_2,\ a_3 \le 2^{a_1}\}$, and matrices (0) identical zero, M is called a <u>masking field</u> iff the following two properties are true :

(i) $\forall a_1,a_2,a_3,a_2',a_3'\ (((a_1,a_2,a_3)\in I_n\ \&\ (a_1,a_2',a_3')\in I_n) \rightarrow M_{a_1a_2a_3'} = (0) \vee M_{a_1a_2'a_3'} = (0) \vee M_{a_1a_2a_3} = M_{a_1a_2'a_3'})$, and

(ii) $\forall i'j,j'\ ((p_{i'j'} = 1\ \&\ p_{i'j} = 0\ ) \rightarrow \forall i\ (p_{ij} = 0\ ))$.

Note that for any masking field M also property

(iii) $\forall i'j,j'\quad ((p_{i'j} = 1\ \&\ p_{i'j'} = 1\ ) \rightarrow \forall i\ (p_{ij} = p_{ij'}))$

is true.

<u>Corollary 3.</u> Let $M\in\mathfrak{M}_n$, $n = 2^m$, $(a_1,a_2,a_3)\in I_n$ and $(a_1,a_2',a_3')\in I_n$. Then, if $M_{a_1a_2a_3} \neq (0)$ and $M_{a_1a_2'a_3'} \neq (0)$ and $M_{a_1a_2'a_3} = (0)$ and $M_{a_1a_2a_3'} = (0)$ it follows that $a_2 = a_2'$ or $a_3 = a_3'$.

This Corollary follows immediately by (i) and (ii). By this Corollary we know, for example, that the right pattern given above is not a masking field.

Now, let $MASK_{2m} := \{(A_{2m-1}A_{2m-2}\ldots A_0):\ A_i\in\{\overline{0},\ \overline{1},\ X\}\ \&\ i=0,1,\ldots,2m-1\}$, for $m\geqslant 1$, and Φ be a mapping of these bit patterns into the set $\mathfrak{M}_n$, $n = 2^m$, with $\Phi((A_{2m-1}A_{2m-2}\ldots A_0)) = (p_{ij})_{i,j=0,1,\ldots,n-1}$ and

$p_{ij} = 1$ iff $i\in\{a_{2m-1}2^{m-1} + a_{2m-2}2^{m-2} + \ldots + a_m 2^0\ :\ a_k\in A_k\ \&\ k=m,\ldots,2m-1\}$, and
$j\in\{a_{m-1}2^{m-1} + a_{m-2}2^{m-2} + \ldots + a_0 2^0:\ a_k \in A_k\ \&\ k=0,1,\ldots,m-1\}$.

Of course, by $\Phi(MASK_{2m})$, $n = 2^m$, the set of all patterns is given which are possible on the array of SQUARE by the used masking scheme. Without proof we state that the following relation is true.

<u>Theorem 4.</u> For $m \geqslant 1$, the set $\Phi(MASK_{2m})$ is equal to the set of all masking fields of size $2^m \times 2^m$.

Using this result, in [3] a decision procedure for the set of all masking fiels of size $2^m \times 2^m$ is given. But, in general the algebraic description of possible patterns by masking fields should be sufficient for decisions about possible structures of patterns by the given masking scheme during the design of PARSIL programs.

3. LOWER TIME BOUNDS BY DATA TRANSFER ANALYSIS

The general goal for writing programs on the parallel simulator should be to go as good as possible to the minimal time bound for the given computational problem on the simulated parallel structure. Because one essential point during the design of PARSIS was to ensure simple exchange of interconnection schemes in [4] a general approach for analyzing lower time bounds by quantitative descriptions of possible data transfer within a concrete interconnection network was outlined. For axample, the local transfer analysis was based on the local transfer function Λ_{SYS}, where $\Lambda_{SYS}(t)$ is defined to be the maximal number of input data in different registers of SYS at the beginning of a computational process on SYS that may have influence on a register of SYS after t steps of computation, where the maximum is taken on all possible input situations, and all possible computational processes on SYS. For example, let SYS be the SQUARE system as descrived above. Then

$$\Lambda_{SQUARE}(t) = 2t^2 + 3t + 1$$

follows, for $t \geq 0$. In [4] this local transfer function was given for about 18 different interconnection structures, making differences between off-line and on-line mode, as well as global transfer function.

Let $n,m \geq 1$ and f be an n-ary function defined on a certain set domain(f) of n-tuples of real numbers, and into the set of m-tuples of real numbers. For an n-tuple $(x_1,x_2,\dots,x_n) \in$ domain(f), define

$$sub_i(x_1,x_2,\dots,x_n) = \{j:\ 1 \leq j \leq n\ \&\ \exists x'(x' \neq x_j\ \&\ (x_1,x_2,\dots,x_{j-1},x',x_{j+1},\dots,x_n) \in domain(f)\ \&\ proj_i(f(x_1,x_2,\dots,x_n)) \neq proj_i(f(x_1,x_2,\dots,x_{j-1},x',x_{j+1},\dots,x_n))\}$$

to be the set of all positions j such that changes in the jth component of $(x_1,x_2,\dots,x_n)$ have an effect on the projection $proj_i f$, for $1 \leq i \leq m$. Then, define

$$\lambda_f = \max_{(x_1,x_2,\dots,x_n)} \ \max_{1 \leq i \leq m} \ card(sub_i(x_1,x_2,\dots,x_n)) .$$

The function f is called locally d-dependent iff $d \leq \lambda_f$. It follows that for arbitrary computational systems SYS the following Theorem holds.

Theorem 5. The computation of function f on SYS requires at least t_o steps of computation in the worst case, where $\Lambda_{SYS}(t_o) \geq \lambda_f$.

In some cases of f the computation of the exact value of λ_f is quite a sophisticated problem but a value of d close to λ_f should be possible in general. For example, for the computational problem of multiplying two N x N matrices it follows immediately that

$\lambda_{MATRIX\ MULTIPLICATION} = 2N$
leading to the lower time bound $t_o = (2N - \frac{7}{16})^{1/2} - \frac{3}{4}$ on SQUARE, analogously for computing the determinant of a N x N matrix the value
$\lambda_{DETERMINANT} = N^2$,
for the 2Dfourier transform of N x N complex matrices the estimate
$2N^2 - 4 \leq \lambda_{2D-DFT} \leq 2N^2 - 1$,
or for computing the convex hull of a simple polygon with N edges the value
$\lambda_{CH\_SIPOL} = 2N$
may be used for obtaining lower time bounds. Similar approaches were described in [1, 2] , a more carefull treatment of the matrix multiplication problem for SQUARE in [2] leads to the lower time bound $t_o = \frac{1}{4}(2N^2 - 1)^{1/2} - \frac{1}{4}$.

4. ALGORITHMS ON SQUARE

For a first demonstration of the flexible work with PARSIS a class of computational problems was selected which are connected with a special set of binary images. For SQUARE this general orientation to the image processing area seems to be quite natural because many parallel algotihms for this field are known till now, and implementation are possible with recent technologies.

Let $\underline{Z}$ be the set of all integers, $\underline{Z}^2$ may be considered as the infinite orthogonal grid for describing binary pictures. Let be CS the set of all finite grid point sets $G \subseteq \underline{Z}^2$ satisfying the following two axioms:

(i) $P \in G \rightarrow card(\eta_8(P) \cap G) \geq 2$, and

(ii) $P \in G \rightarrow \exists Q(Q \in G - \{P\}\ \&\ card(\eta_8(Q) \cap (G - \{P\})) \leq 1)$,

where $\eta_8(P)$ denotes the 8-neighborhood of point P ("around P"). In CS we have collected all <u>cellelur structures</u> in $\underline{Z}^2$. During the first phase of implementation of PARSIS different programs for SQUARE were written connected with the set CS.

For $G \in CS$ let $G_1^{int}, G_2^{int}, \ldots, G_r^{int}$ be the maximal 4-connected (i.e. connected with respect to the 4-neighborhood components of $\overline{G} = \underline{Z}^2 - G$ of finite cardinality. Let G_i be the union of G_i^{int} with the set of all points in G which are 4-neighbors of points in the set G_i^{int} , for i= 1,2,...,r. $G_1, G_2, \ldots, G_r$ are the <u>cells</u> of the cellular structure G with interiors G_i^{int} and borders $G_i^{border} = G_i - G_i^{int}$. Let G^{int} be the union of all sets G_i^{int} and G^{border} be the union of all sets G_i^{border}. Finally the representing point of cell G_i is that point of G_i^{int} with minimal x-coordinate, and within the set of all points of G_i^{int} with minimal x-coordinate that with minimal y-coordinate. By these notions some computational problems may be formulated which have practical importance in digital binary image analysis.

A cellular structure G is represented by a binary picture f if $(x,y) \in G$ iff $f(x,y) = 1$. For example, the following binary picture represents a cellular structure, the representing points of the cells are marked by ■ .

```
0 0 1 1 1 1 0 0 0 1 1 1 1 1 1 0
0 1 ■ 0 0 0 1 0 1 ■ 0 0 0 0 0 1
1 0 0 0 0 0 1 0 0 1 0 0 0 0 0 1
1 0 0 0 0 1 0 0 0 0 1 1 1 1 1 0
0 1 0 1 1 0 0 0 1 0 0 0 0 0 0 0
0 0 1 ■ 0 1 1 1 0 1 1 0 0 0 0 0
0 1 0 0 0 0 0 0 0 0 0 1 0 0 0 0
1 0 0 1 0 0 0 0 0 0 1 0 0 1 1 0
1 0 1 0 1 1 1 0 0 1 ■ 1 1 0 0 1
0 1 0 0 0 0 0 1 1 0 0 0 0 0 0 1
0 0 0 0 0 0 0 0 0 1 0 0 0 0 1 0
0 0 0 0 0 0 0 0 0 0 1 1 1 1 0 0
0 0 1 1 1 1 1 1 1 1 ■ 0 0 0 1 0
0 1 0 0 0 0 0 0 0 0 0 0 1 1 0 0
0 1 0 0 0 1 1 0 0 0 0 1 0 0 0 0
0 0 1 1 1 0 0 1 1 1 1 0 0 0 0 0
```

In this picture, G consists of two 8-connected components and five cells.

At first, the transformation of arbitrary binary pictures into pictures representing cellular structures by thinning and cuting off "arms" was sinsidered. For example, if T the maximal thickness of areas of points with value 1 in the input picture measured on the base of 4-neighborhood then the first step, thinning of binary images, may be done in time $8 2 \cdot T + 5$ on SQUARE.

For analyzing cellular structures some examples of algorithms and the time measured by the number of basic instructions on SQUARE are as follows:

1. Marking all representing points of the cells by a special value not equal to 0 or 1 in time $42 \cdot D + 30$ where D denotes the maximal distance of a point $P \in G_1^{int}$ to the representing point of G_1 in the 4-metric;

2. Marking all points of G_1^{int} with its distance to the representing point of G_1 in the 4-metric in time $19 \cdot D + 25$ where D as above

3. Coloring of calls with the maximal value which a point of this cell has at the beginning in time $20 \cdot D + 19$ where D as above;

4. Numbering of representing points of the cells in a unique way by numbers 1,2,...,r in time $21 \cdot (N-2) + 15$, where $N \times N$ is the size of the input picture;

5. Computing the Euler numer of the cellular structure [8] in time $14 \cdot (N-1) + 14$ where N as above;

6. Marking maximal squares within sets G_1^{int} with one side parallel to the diagonal of the $N \times N$ input picture in time $2 \cdot M^2 + 15 \cdot M + 17$ where M is 1 plus the maximal half length of the diagonals of these squares;

7. Computing the smallest rectangles with sides parallel to the coordinate axes containing the cells G_i in time $70 \cdot M + 37$ where M denotes the maximal length of one of the sides of these rectangles;

8. Deciding whether two points belong to the same cell or not in time $6 \cdot M + 21$ where M is the 4-distance between these two points.

Detailed descriptions of these, and further algorithms on SQUARE are given in [3]. For example, 90° rotation of a N x N input picture may be done on SQUARE in time $12 \cdot N + 8 \cdot \log_2 N - 78$, and similar results are obtained for reflections at the diagonal or at the middle axes. Some of the constants mentioned above may be made available at the beginning of computation, as N and M in 8., e.g., the others may be selected a-priori for a given class of pictures of cellular structures. By comparing the time results with lower time bounds by data transfer analysis (what is just one approach to charakterize lower time bounds) the algorithms in points 1 to 8 are known to be close to the optimum, for example $\lambda_{EULER-NUMBER} = N^2$ leading to $t_o = (N^2 - \frac{7}{16})^{1/2} - \frac{3}{4}$.

References

[1] Abelson, H.: Lower bounds on information transfer in distributed computations, J. ACM 27 (1980), 384 - 392.

[2] Gentleman, W.M.: Some complexity results for matrix computation on parallel processor, J. ACM 25 (1968), 122 - 115.

[3] Heinrich, R.: Parallele Binärbildanalyse mittels PARSIS-Simulator. Friedrich-Schiller-Universität Jena, Sektion Mathematik, Juni 1984.

[4] Klette, R.: Analysis of data flow for SIMD systems. University of Maryland at College Park, Center for Automation Research, CAR-TR-1, March 1983 (to appear: Acta Cybernetica).

[5] Klette, R., Heinrich, R., Wedler, V.: Konzeption und Realisierungsstand eines Parallelsimulationssystems. TU Braunschweig, Tagungsband Zellulartreffen Februar 1984, Institut für Informatik.

[6] Legendi, T.: A cellular processor project. Int. Workshop on Parallel Processing by Cellular Automata, Berlin, GDR, Sept. 15 - 16, 1982.

[7] Siegel, H.J.: A model of SIMD machines and a comparison of various interconnection networks. IEEE Trans. Computers C-28 (1979), 907 - 917.

[8] Vollmar, R.: Algorithmen in Zellularautomaten. Teubner Studienbücher Informatik, Stuttgart 1979.

A FUNCTIONAL DESCRIPTION OF SYNCHRONOUS SYSTEMS

Christoph Meinel +

Introduction:

Analysing and controlling the data choreography in synchronous parallel systems is one of the most striking problems with handling such systems. In the following paper an attempt is made to manage this problem within a mathematical framework. The essential point of our functionally describing approach is the introduction of delayoperators for describing the work of the globally clocked registers beside of the combinatorial functions describing the work of the combinational logic elements. Thereby the computational complexity of this functional elements will be not restricted what makes it possible to model electronic circuits as well as whole parallel machines. With the help of these mathematical operators and functions the value can be computed leaving a certain point of the system in a certain clock period whenever the input data which have arrived the system up to that clock period are known. Thus we are able for instance to give mathematical correctness proofs for design proposals of synchronous parallel systems. Likewise we can derive rules for rearranging synchronous systems for providing more efficient systems to a given one within our mathematical framework because we are able to decide the computational equivalence of different systems.

In the first paragraph our graph theoretic model of a synchronous system will be sketched. The second paragraph is concerned with the functional description of such systems. For illustration functional descriptions of a design of a VLSI array for linear time GCD computation will be derived. In the last paragraph some methods for optimizing synchronous systems will be mentioned.

+ Sektion Mathematik, Humboldt-Universität, DDR-1086 Berlin, PSF 1297

1. The graph-theoretic model of a synchronous parallel system

A synchronous parallel system can be regarded as a system which is partitioned into input ports, functional elements (combinational logik) and globally clocked registers (clocked memories). Its interconnection pattern can be described by a map

$$m = (V, X, d)$$

which consists of

- a set V of points,
- a finite set X of (directed) connections and
- a partial function d

$$d: V \times X \xrightarrow{\supset} V$$

which assigns a new point $vx := d(v,x)$ to a point $v \in V$ via a connection $x \in X$.

Let us remark that the function d determines a function val

$$\begin{aligned} val: V &\longrightarrow 2^V \\ v &\longmapsto val(v) := \{x \mid vx \text{ is defined}\} \end{aligned}$$

wich assigns to a point v the set of all connections originated in v. Because the names $x \in X$ of the directions are used only to order the connections origianted in a single point we take for X the set $[n] := \{0,1,\ldots,n\}$ when $|X| = n$ such that the name of a connection stands for the number of this input channel. In order to modelize synchronous systems we restrict us to maps satisfying the following two conditions:

(1) For all points $v \in V$ it holds either
$val(v) = \emptyset$ or $val(v) = \{0\}$ or $val(v) = \{1,2,\ldots i\}$
for an $i \in [n] - \{0\}$.

(2) If $vp := d^*(v,p) := d(d^*(v,p_1 \ldots p_{s-1}), p_s) = v$ for a $v \in V$,
$p = p_1 \ldots p_s \in [n]^*$ then $p \notin \{1,\ldots,n\}^*$.

Thereby, the second condition guarantees that every cycle of the map contains at least one connection labelled by 0. Maps satisfying condition (1) and (2) are called communication maps.

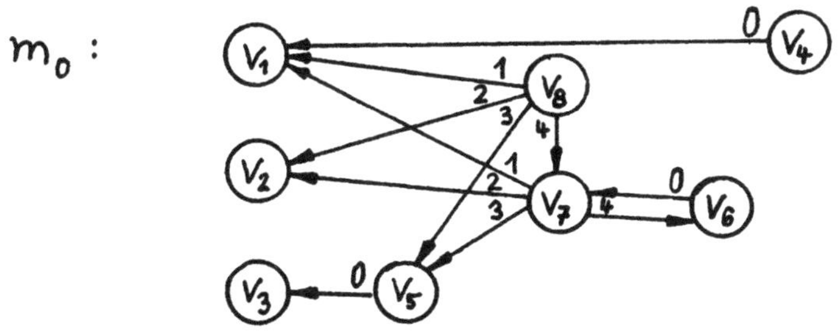

Fig. 1 Example of a map

Obviously, this map is a communication map because the only cycle $v_7 = v_7 30$ or $v_6 = v_6 03$, respectively, contains a connection labelled by 0, and because the valency sets of all points are of the special forms predicted in condition (1).

Via a function l which is called a legend we assign to the vertices of a communication map the mathematical descriptions of the computational behavior of the electronic devices (regarded as black boxes) which correspond to the regarded vertices and which communicate along the connections of the map. Thereby we interprete the directions of the connections as the demand for communication. In more detail:

If a point v corresponds to an input port, i.e. if $val(v) = \emptyset$ then l assigns to v a formal input variable from the set $\{x_j\}_{j \in \mathbb{N}}$.

If A is the input and communication alphabet and if a vertex v corresponds to a combinational logic element, i.e. if $|val(v)| > 0$ and $val(v) = \{1,\ldots,|val(v)|\}$ then l assigns to v the function F

$$F : A^{|val(v)|} \longrightarrow A$$

which describes the input-output behavior of this combinational logic element. In different contexts this combinational logic element might be of different computational complexity such as NAND or NOR gates, multiplexors or RAM's, for example.

Finally, if v corresponds to a register, i.e. if $val(v) = \{0\}$ then l assigns to v a delay operator R_c, $c \in A$ which is defined on infinite sequences $(a^{(i)})_{i \geq 1} \in A^{\mathbb{N}}$ by

$$R_c(a^{(i)}) := \begin{cases} a^{(i-1)} & \text{if } i > 1 \\ c & \text{otherwise} \end{cases}$$

with the initial value $c \in A$ of the register corresponding to v. These delay operators reflect the properties of the globally clocked registers which are assumed to have a single input (via its only connection labelled by 0) and a single output, and which make the input sampled in the $(i-1)$-th clock period available in the i-th clock period as an output.

To give an example let us take a finite field A as the communication alphabet and define a legend l_o to the communication map m_o of figure 1 by

$$l_o(v_i) = x_i \quad \text{for } i = 1,2,3,$$

$$l_o(v_4) = l_o(v_6) = R_0 \quad \text{and} \quad l_o(v_5) = R_1,$$

$$l_o(v_8) = c : \quad A^4 \longrightarrow A$$

$$(a_1,a_2,a_3,a_4) \longmapsto \begin{cases} 0 & \text{if } a_3 \neq 0 \\ a_1 - a_2/a_4 & \text{otherwise} \end{cases}$$

and

$$l_0(v_7) = q : \quad A^4 \longrightarrow A$$
$$(a_1,a_2,a_3,a_4) \longmapsto \begin{cases} a_1/a_2 & \text{if } a_3 \neq 0 \\ a_4 & \text{otherwise} \end{cases} .$$

Now the importance of the two conditions (1) and (2) making a map to a communication map can be recognized easely. Condition (1) guarantees the right joining conditions for the functional elements and the registers and condition (2) serves as avoid uncontrollable ozcillation. The pair $\mathcal{f} = (m,l)$ of a communication map m and a legend l will be called a synchronous (parallel) system.

2. Functional description of the computational behavior of synchronous systems

An synchronous system $\mathcal{f} = ((V,\{0,\ldots,n\},d),\ l)$ starts its computations when it gets input data via the input vertices from the external world. Let us assume that the system gets in the i-th clock period the input $x_j^{(i)} \in A$ via an input vertex labelled by X_j. (Sometimes it is more convenient to think of an input vertex as of an infinite chain of registers every of which contains the according input value.) These input signals and the signals output by the (globally clocked) registers proceed in any clock period along all possible connections through the functional elements which perform their computations up to the next registers. For describing the output behavior of a certain point $v \in V$ of the system we have to determine all informations (elements of A) and all combinatorial functions of the system which are of influence on the output of v produced in the i-th clock period. To do this we have to regard the accessibility component $\mathcal{f}_v$ of v in $\mathcal{f}$ which is the full subsystem of $\mathcal{f}$ containing all points $v' \in V$ which are connected with v in $\mathcal{f}$ via a path $p = p_1 \ldots p_s \in \{0,1,\ldots,n\}^*$

$$v' = vp .$$

The signals moving in this subsystem $\mathcal{f}_v$ are exactly those data which are of influence on the data produced in v during the different clock periods. In order to compute these data let us denote the datum which will be computed in the vertex $v \in m$ and left v in the i-th clock period by $v^{(i)} \in A$.

If v is an input vertex with $l(v) = X_j$ then the input $x_j^{(i)}$ left v in the i-th clock period, i.e.

$$v^{(i)} = x_j^{(i)} .$$

Otherwise it holds

$$v^{(i)} = l(v)(\ (vx)^{(i)}\ |\ x \in val(v)) , \qquad (2.1)$$

or in more detail:

If $l(v) = R_a$ for an $a \in A$ and, therefore, if $val(v) = \{0\}$ then

$$v^{(i)} = R_a(v0)^{(i)} = \begin{cases} (v0)^{(i-1)} & \text{if } i > 1 \\ (v0)^{(0)} := a & \text{otherwise} \end{cases},$$

and if $l(v) = F$ and, therefore, if $F : A^{|val(v)|} \longrightarrow A$ then

$$v^{(i)} = F((v1)^{(i)}, \dots, (v|val(v)|)^{(i)}) .$$

To illustrate these formulas let us regard the synchronous system $\mathcal{f}_0$ defined in the previous section, and compute the output behavior of its points:

$$v_j^{(i)} = x_j^{(i)} \quad \text{for } j = 1,2,3 ,$$

$$v_4^{(i)} = R_0(v_4 0)^{(i)} = v_1^{(i-1)} = \begin{cases} 0 & \text{if } i-1 = 0 \\ x_1^{(i-1)} & \text{otherwise} \end{cases},$$

$$v_5^{(i)} = R_1(v_5 0)^{(i)} = v_3^{(i-1)} = \begin{cases} 1 & \text{if } i-1 = 0 \\ x_3^{(i-1)} & \text{otherwise} \end{cases},$$

$$v_6^{(i)} = R_0(v_7)^{(i)} = v_7^{(i-1)} ,$$

$$v_7^{(i)} = q(v_1^{(i)}, v_2^{(i)}, v_5^{(i)}, v_6^{(i)}) = q(x_1^{(i)}, x_2^{(i)}, x_3^{(i-1)}, v_6^{(i)}) ,$$

$$v_8^{(i)} = c(v_1^{(i)}, v_2^{(i)}, v_5^{(i)}, v_7^{(i)}) = c(x_1^{(i)}, x_2^{(i)}, x_3^{(i-1)}, v_7^{(i-1)}) .$$

It is clear that cycles in a synchronous system give rise to recursions in the computations which are solvable because the communication maps underlying the regarded synchronous systems satisfy condition (2).
For our example one obtains, successively,

$$v_6^{(1)} = 0 ,$$

$$v_7^{(1)} = q(x_1^{(1)}, x_2^{(1)}, 1, 0) = x_1^{(1)}/ x_2^{(1)} ,$$

$$v_8^{(1)} = c(x_1^{(1)}, x_1^{(1)}, 1, x_1^{(1)}/ x_2^{(1)}) = 0 ,$$

and

$$v_6^{(2)} = v_7^{(1)} = x_1^{(1)}/ x_2^{(1)} ,$$

$$v_7^{(2)} = \begin{cases} x_1^{(1)}/ x_2^{(1)} & \text{if } x_3^{(1)} = 0 \\ x_1^{(2)}/ x_2^{(2)} & \text{otherwise} \end{cases},$$

$$v_8^{(2)} = \begin{cases} 0 & \text{if } x_3^{(1)} \neq 0 \\ x_1^{(2)} - x_2^{(2)}(x_1^{(1)}/ x_2^{(1)}) & \text{otherwise} \end{cases},$$

and, generally,

$$v_6^{(i)} = \begin{cases} x_1^{(k)}/ x_2^{(k)} & \text{if } x_3^{(k)} \neq 0 \text{ and } x_3^{(j)} = 0 \\ & \text{for all } j ,\ k < j \leq i-1 \\ x_1^{(i-1)}/ x_2^{(i-1)} & \text{if } x_3^{(i-1)} \neq 0 \end{cases},$$

$$v_7^{(i)} = \begin{cases} x_1^{(k)} / x_2^{(k)} & \text{if } x_3^{(k)} \neq 0 \text{ and } x_3^{(j)} = 0 \text{ for all } j,\ k < j \leq i-1 \\ x_1^{(i)} / x_2^{(i)} & \text{if } x_3^{(i-1)} \neq 0 \end{cases},$$

$$v_8^{(i)} = \begin{cases} 0 & \text{if } x_3^{(i-1)} \neq 0 \\ x_1^{(i)} - x_2^{(i)}(x_1^{(k)} / x_2^{(k)}) & \text{if } x_3^{(k)} \neq 0 \text{ and } x_3^{(j)} = 0 \text{ for all } j,\ k < j \leq i-1 \end{cases}.$$

(Observing our system $\mathcal{f}_0$ carefully one recognizes that it can be used to compute the greatest common divisor (GCD) of two polynomials with coefficients in the finite field A . Indeed, the given system models one stage of a systolic VLSI array for a linear time GCD computation for polynomials proposed by Brent/Kung /1/. This design makes the most of the following fact:

$$GCD(P,Q) = GCD(\overline{P},Q)$$

with

$$\overline{P} = \overline{P}(P,Q) = P - (p^{(1)}/q^{(1)}) * t^{(\deg P - \deg Q)} * Q$$

for

$$P = p^{(1)} * t^r + \ldots + p^{(r+1)}$$

and

$$Q = q^{(1)} * t^s + \ldots + q^{(s+1)} \quad \text{and} \quad r \geq s .$$

Indeed, one can see that $\mathcal{f}_0$ generates the desired polynomial $\overline{P}$ if one inputs the coefficients $p^{(i)}$, $q^{(i)}$ via the input ports v_1 and v_2 , respectively, and if the system gets via the input port v_3 a sequence of 0's and 1's which signalizes the position of the leading coefficient of two consecutive polynomials. Moreover, if one takes a lot of copies of $\mathcal{f}_0$ and arrange these copies in a linear array by identifying the vertices

v_4 and v_1
v_8 and v_2
v_5 and v_3

of two consecutive copies (and providing the system with a facility for exchanging the roles of the two polynomials if $\deg \overline{P} < \deg Q$ which we omitted to simplify matters) then one gets a parallel working system for linear time GCD computation because any copy of $\mathcal{f}_0$ provides the generated pair $\overline{P}$, Q in the right configuration for generating the pair $\overline{\overline{P}}$, Q or $\overline{P}$, $\overline{Q}$, respectively, in the consecutive copy of $\mathcal{f}_0$. A formal correctness proof for this design which was not given in /1/ can be given easily with the help of our functional description by computing the output behavior of the output ports under consideration in the manner described above.)

Instead of computing the values $v^{(i)}$ which left the point v of the synchronous system $\mathcal{f} = (m,l)$ in the i-th clock period from the values leaving the points vx, $x \in val(v)$ we can compute them from the values leaving the points vp of $\mathcal{S}$ where p is runing over a complete set of initial paths P of $\mathcal{S}_v$ which contains exactly one initial segment of all sufficient long path of $\mathcal{S}_v$ starting in v. For example, the following sets are such complete sets of initial paths of $\mathcal{S}_v$:

$$P_v^1 := val(v) ,$$

$$P_v^k := \{p \mid p \in \{0,1,\dots,n\}^* , vp \text{ is defined and } |p| = k\}$$

$$R_v^k := \{p \mid p \in \{0,1,\dots,n\}^* , vp \text{ is defined and } \varrho_v(p) = k \text{ and } l(vp) = R_a \text{ for an } a \in A\}$$

for all k .(ϱ_v is a monoid homomorphism $\varrho_v : \{0,1,\dots,n\}^* \longrightarrow \mathbb{N}$ which counts the edges of p labelled by 0 and therefore the number of registers on p .) Applying formula (2.1) successively we obtains for such a complete set P of initial values the formula

$$v^{(i)} = f_p((vp)^{(i)} \mid p \in P) \tag{2.2}$$

where f_p is built from the combinational functions and the delay operators assigned via the legend l to the points "between" v and vp for all $p \in P$. Further, if we apply the delay operators R_a , $a \in A$ whenever this is possible then we can compute $v^{(i)}$ as the value of a purely combinatorial function e_P

$$v^{(i)} = e_P((vp)^{(i-\varrho_v(p))} \mid p \in P) \tag{2.3}$$

because the number of registers $\varrho_v(p)$ along p equals the number of the applyments of the delay operators along this path p .
The last formula (2.3) gives rise to the following statement which will be formulated as a lemma because of its importance.

<u>LEMMA:</u> The data $v^{(i)}$ computed in a synchronous system $\mathcal{S}$ in the i-th clock period in the point v is determined uniquely by the values stored in the $(i-j)$-th clock period in the registers located in points $v' = vp$ of R-distance j (i.e. $\varrho_v(p) = j$) for each $j < i$.

<u>PROOF:</u> Taking R_v^j as the complete set of initial values one obtains from formula (2.3)

$$v^{(i)} = e_{R_v^j} ((vp)^{(i-j)} \mid p \in R_v^j)$$

what proves our lemma. ☒

Let us write e_v^j for the combinatorial function $e_{R_v^j}$, easily, to simplify notations.
Finally let us define what we will understand by speaking of computational equivalent systems. To that end let $\mathcal{S}$ and $\mathcal{S}'$ be two synchronous systems with the input vertices $v_1,\dots,v_k$ and $v_1',\dots,v_k'$.

respectively, $\mathcal{S}$ and $\mathcal{S}'$ are called <u>computational equivalent in the points $v \in \mathcal{S}$ and $v' \in \mathcal{S}'$</u> if the output behaviors of these points after an initial working phase are identically under identical input conditions. That means formally there is a permutation $\pi \in \mathcal{T}_k$ such that from $v_1^{(i)} = v'^{(i)}_{\pi(1)}, \ldots, v_k^{(i)} = v'^{(i)}_{\pi(k)}$ for all $i \in \mathbb{N}$ it follows

$$v^{(i)} = v'^{(i)} \quad \text{for all} \quad i \geq i_0 \; , \; i_0 \in \mathbb{N} \, .$$

(We do not require that it holds $v^{(i)} = v'^{(i)}$ for all $i \in \mathbb{N}$ because it is sometimes impossible to initialize two systems providing in some points the same outputs from a certain clock period on in such a way that they produce the same outputs in all clock periods. However, it is useful to regard such systems as computational equivalent in these points, too.)

3. Optimizing synchronous parallel systems

Beside of giving correctness proofs for design proposals for synchronous systems our functional description of such parallel systems has been proved to be very useful in deriving rules for rearranging such systems in order to provide more efficient systems. To give an example of such a rearrangement let us regard the following exchange rule which says, that in a synchronous system the following subsystems

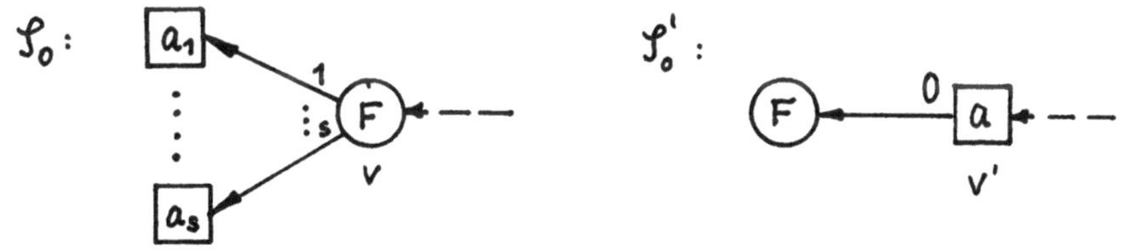

Fig. 2 Computational equivalent subsystems

can be replaced one by the other without essential changes in the computational behavior under certain conditions. Such a relocation of the registers within a synchronous system which will be reflected in the functional description of the system by commuting the delay operators with the combinatorial functions can reduce combinational rippling and, therefore, lead to a shorter clock period without any increase in circuit latency. This optimization technique for synchronous systems was investigated explicitely first by Leiserson/Saxe in /3/. But because of a lack of a exact mathematical description of their systolic systems they could not formulate and prove exactly their retiming lemma which states that such a relocation of registers does not changes the computational behavior of the system.

Before we are going to prove a revised version of this lemma we have to define some notions. Let $\mathcal{S} = (m,l)$ be a synchronous system with the communication map $m = (V,X,d)$. A synchronous system $\mathcal{S}' = (m',l')$ with the communication map $m' = (V',X',d')$ is called a <u>subsystem</u> of $\mathcal{S}$

$$\mathcal{S}' \subseteq \mathcal{S}$$

if it holds

$$V' \subseteq V, \quad X' \subseteq X,$$

$$val'(v) = val(v) \text{ or } \emptyset,$$

$$d'|_{V' - \{v \mid val'(v) = \emptyset\}} = d|_{V' - \{v \mid val'(v) = \emptyset\}'}$$

and

$$l'|_{V' - \{v \mid val'(v) = \emptyset\}} = l|_{V' - \{v \mid val'(v) = \emptyset\}}.$$

Thereby a point $v' \in V'$ is called an <u>exit</u> of $\mathcal{S}'$ in $\mathcal{S}$ if there exists a $v \in V - V'$ and a $x \in X$ with $vx = v'$ in $\mathcal{S}$. Now we are able to formulate our first exchange theorem.

<u>EXCHANGE THEOREM:</u> Let $\mathcal{S}'$ be a subsystem of the synchronous system $\mathcal{S}$ which satisfies the property that no exit of $\mathcal{S}'$ in $\mathcal{S}$ is contained in a cycle of $\mathcal{S}$ (i.e. a infinite path over a finite carrier). Further let $\mathcal{S}''$ be a synchronous system that is computational equivalent to $\mathcal{S}'$ in all exits of $\mathcal{S}'$ in $\mathcal{S}$. Then the synchronous system $\overline{\mathcal{S}}$ which is built from the system $\mathcal{S}$ by a suitable replacement of $\mathcal{S}'$ by $\mathcal{S}''$ is computational equivalent in all points remaining unchanged to the system $\mathcal{S}$.

For a <u>PROOF</u> and some more details see /4/.

Now it is easy to obtain the retiming lemma for acyclic and finite systems as a corollary of the exchange theorem.

<u>RETIMING LEMMA</u> for acyclic and finite systems: If a synchronous system $\mathcal{S}$ contains one of the two systems $\mathcal{S}_o$ or $\mathcal{S}'$ of figure 2 as a subsystem with the exit v or v', respectively, and if this exit is not contained in a cycle of $\mathcal{S}$ then one of these systems can be exchanged by the other without changing the computational behavior of the whole system in all other points apart from some unessential changes in a starting phase.

<u>PROOF:</u> By computing the output sequences of v and v' of $\mathcal{S}_o$ and $\mathcal{S}_o'$ it can be seen that these two systems are computational equivalent:

$$v^{(1)} = F(R_{a_1}x_1^{(1)}, \ldots, R_{a_s}x_s^{(1)}) \qquad v'^{(1)} = R_aF^{(1)}$$

$$= F(a_1, \ldots, a_s) \qquad\qquad = a$$

$$v^{(2)} = F(R_{a_1}x_1^{(2)}, \ldots, R_{a_s}x_s^{(2)}) \qquad v'^{(2)} = R_aF^{(2)} = F^{(1)}$$

$$= F(x_1^{(1)}, \ldots, x_s^{(1)}) \qquad\qquad = F(x_1^{(1)}, \ldots, x_s^{(1)})$$

and

$$v^{(i)} = F(x_1^{(i-1)}, \ldots, x_s^{(i-1)}) \qquad v'^{(i)} = F(x_1^{(i-1)}, \ldots, x_s^{(i-1)})$$

that means $v^{(i)} = v'^{(i)}$ for all $i > 1$.

Because of this computational equivalence of $\mathcal{S}_o$ and $\mathcal{S}_o'$ it follows from the exchange theorem that an exchange of $\mathcal{S}_o$ and $\mathcal{S}_o'$ in $\mathcal{S}$ does not influence the computational behavior in the unchanged points of $\mathcal{f}$. ☒

Because an exchange of $\mathcal{f}_u$ by $\mathcal{f}'_u$ and vice versa in a synchronous system $\mathcal{f}$ will be reflected in the functional description of the system by an exchange of the delay operators and the combinatorial functions we can restate the retiming lemma for acyclic systems.

<u>CORLLARY:</u> Two acyclic and finite synchronous systems are computational equivalent if their functional descriptions can be made identical by applying suitably the following commutativity rule

$$R_a F(x_1^{(i)},\dots, x_s^{(i)}) = F(R_{a_1} x_1^{(i)},\dots, R_{a_s} x_s^{(i)})$$

for any combinatorial function F and any $a_1,\dots, a_s, a \in A$.

The situation of cyclic systems is much more complicated because the initial values stored in the registers are of a permanent influence on the computations of the system in contrary to the acyclic case in which these values are only of influence in a starting phase of the system. Nevertheless, there can be proved an exchange theorem for cyclic synchronous systems, too, which has to take more care on the initial situation of the system, see /4/. Especially, a retiming of a cyclic system is possible only under very restricted conditions.
Finally let us remark that retiming is only a very restricted exchange rule. A great variety of rules providing systems meeting special requirements and systems with better efficiency coefficients can be derived from our exchange theorems. Some examples can be found again in /4/ .

<u>REFERENCES:</u>

/1/ Brent,R.P., H.T.Kung: Systolic VLSI array for linear-time GCD computation. Proc. of the VLSI Conf., Trondheim, Norway, 1983

/2/ Huffman, D.A.:The synthesis of linear sequencial coding networks.

/3/ Leiserson, Ch.E., J.B.Saxe: Optimizing synchronous systems. MIT/LCS/TM-215, 1982

/4/ Meinel, Ch.: A functional description of synchronous systems. In preparation for publication, 1984

/5/ Meinel, Ch.: Structural investigations of parallel computing devices. Computers and artificial intelligence No. 4, 1984

STRUCTURAL AND BEHAVIORAL EQUIVALENCE RELATIONS IN AUTOMATA NETWORKS

T.Saito and H.Nishio

1. Introduction

The automata network is an information processing system consisting of intercommunicating finite automata. We can interpret it not only as abstraction of biological systems (like a neural network), but also as theoretical framework of artificial information processing systems (like a logical network). From the logical point of view, the automata network is a generalization of cellular automaton.[1] The cellular automaton is uniform both in structure and element automaton. The theory of automata networks can be considered to be the structural theory of automata in contrast to the abstract theory of automata.

We formulate the automata network as an edge- and vertex-labeled directed graph. A vertex-label corresponds to the finite automaton which is to be placed at the vertex and an edge-label corresponds to the label of the input terminal of a finite automaton on the vertex.

In this paper, we define, as to the set of vertices, the structural equivalence relation induced by the structure of the graph and the behavioral equivalence relation induced by the behavior (i.e. state transition) of finite automaton on each vertex and investigate the relationships between these relations. We have obtained some results in particular about the element-uniform system. Since the space is limited, we show propositions and theorems without proofs except for a few important results.

2. Definition of Automata Networks

We define the automaton networks in the following way. First we need the notion of a graph.

<u>Definition 2.1</u> Edge-labeled directed graph G

An edge-labeled graph (simply a graph) G is defined by $G = (V',E)$, where V' is a set of finite or infinite number of vertices with conditions $V' = V \cup V_I$, $V \cap V_I = \phi$ and $V \neq \phi$, where V is called the set of inner vertices and V_I the set of input vertices. E is a set of labeled edges, i.e. $E = \{E_1, E_2, \ldots, E_k\}$ where $E_i \subsetneq V \times V'$ and for every vertex $u \in V$ there is at most one vertex $v \in V'$ such that $(u,v) \in E_i$ for every i $(1 \leq i \leq k)$.

We also write $E_i(u) = v$ if $(u,v) \in E_i$, which means that there is an edge labeled with i from v to u.

Department of Biophysics, Faculty of Science, Kyoto University, kyoto, Japan

When we consider the automata network with output, we specify the set of output vertices $V_O \subsetneq V$. If $V_I = \phi$ (i.e. $V' = V$) then the graph G becomes the base of an autonomous automata network.

<u>Definition 2.2</u> Group graph (Uniform graph)

A graph $G = (V,E)$ is a group graph[2], denoted by (V,N), if V is a group with binary operation $\bullet$, and for every $u \in V$ and for every i ($1 \leq i \leq k$) $E_i(u) = e_i \bullet u$ where $e_i \in V$ and $N = \{e_1, e_2, \ldots, e_k\} \subset V$. N is called the neighborhood stencil after the cellular automata theory.

In the following the group operation symbol $\bullet$ will not be written explicitly. If a group graph $G = (G,N)$ is strongly connected, then every vertex u can be expressed by using elements of N as follows; $u = e_i \ldots e_i \lambda$ where $e_i \in N$ and λ is the identity of V.

For example, 2-dimensional von Neumann cellular space is based on the group graph $G = (V,N)$ where $V = \mathbf{Z} \times \mathbf{Z}$, the direct product of the additive group of integers, and $N = \{(0,0),(1,0),(0,1),(-1,0),(0,-1)\}$.

In the case of group graph we assume that an identical automaton is placed at every vertex. That is, we treat a cellular automaton.

<u>Definition 2.3</u> Cell M

An automata network is composed of many cells, each of which is a finite automaton placed at an inner vertex of graph G and defined as follows.

A cell (or element automaton) M is defined by 3-tuple $M = (Q,k,f)$, where Q is a finite set of inner states (Q designates also the set of input and output alphabets), k is the number of input terminals of M and f is the state transition function called the local map defined as $f : Q^{k+1} \rightarrow Q$. If $f(q,x_1,x_2,\ldots,x_k) = q'$, then q' represents the next state provided that the present state is $q \in Q$ and the input to the i-th terminal is $x_i \in Q$. The output of M is defined to be the same as the present state of M. The set of all kinds of cells is denoted by $\mathbf{M}$.

<u>Definition 2.4</u> Automata network A

Suppose that a graph G is given. Then we can define an automata network $A = (G,\alpha)$ by allocating cells to vertices.

That is, in $A = (G,\alpha)$, G is an edge-labeled directed graph and α is an allocation map of cells, i.e. $\alpha: V \rightarrow \mathbf{M}$. If the indegree of $v \in V$ is k, then $\alpha(v)$ is taken to be a k-input cell. It is also assumed that α maps every vertex to a cell having a common state set Q.

$A = (G,\alpha)$ is called the element-uniform automata network if α maps all vertices to one cell. In this case, α is also called uniform. The so called "cellular automaton" is an element-uniform automata network with the group graph.

Now we define the behavior of automata networks after that of cellular automata.

<u>Definition 2.5</u> Behavior of automata networks

Assume that $A = (G,\alpha)$ is given. When f is the local map of $\alpha(v)$ (i.e. the cell M placed at $v \in V$), we sometimes represent it by $\alpha(v)$ if there is no confusion.

Let $\Phi = \{\varphi | \varphi: V_I \rightarrow Q\}$ be the set of all input configurations to automata network A. That is, $\varphi(v) = x$ represents that the input state of the input vertex v of automata network A is x.

$C = \{c | c : V \rightarrow Q\}$ is the set of all state configurations of A as in the case of a usual cellular automaton. We define the next state configuration c' as follows: Let c and φ be the present state configuration and the input configuration to A respectively. For $v \in V$, $c'(v) = \mathcal{T}_\varphi(c)(v) = \alpha(v)(c(v),c(E_1(v)),\ldots,c(E_k(v)))$. We consider that if $E_i(v) \in V_I$ then $c(E_i(v)) = \varphi(E_i(v))$. The mapping $\mathcal{T}_\varphi$: $C \rightarrow C$ thus defined is called the global map as opposed to the local map.

3. Equivalence Relations on V

<u>Definition 3.1</u> Equivalence relations on V

When $G = (V',E)$ is given, we denote by $\mathbf{R}_G$ the set of all equivalence relations (or partitions) on V. For $R \in \mathbf{R}_G$, when the vertices u and v are equivalent in the sense of R, we denote $(u,v) \in R$. $R_0 \in \mathbf{R}_G$ denotes the equivalence relation such that for all pair of vertices u and v $(u \neq v)$ $(u,v) \notin R_0$. That is every pair of vertices are not equivalent. R_U denotes the equivalence relation such that for all $u, v \in V$, $(u,v) \in R_U$. This means that the number of equivalence classes is one. If there is no confusion, we may use R instead of $\mathbf{R}_G$.

When V is finite, an equivalence relation $R \in \mathbf{R}$ is expressed by specifying every equivalence classes like $R = (\{v^1{}_1,\ldots,v^1{}_{c1}\},\{v^2{}_1,\ldots,v^2{}_{c2}\},\ldots)$. So when $V = \{v_1,\ldots v_n\}$, $R_0 = (\{v_1\},\{v_2\},\ldots,\{v_n\})$ and $R_U = (\{v_1,v_2,\ldots,v_n\})$. Sometimes we use the abbreviated notation like $(\{v_1,v_2\},v_3,v_4)$ in place of $(\{v_1,v_2\},\{v_3\},\{v_4\})$.

<u>Definition 3.2</u> Structural equivalence relation (SER)

$R \in \mathbf{R}_G$ is called a structural equivalence relation (SER) of automata network $A = (G,\alpha)$, iff the following conditions are satisfied.

If $(u,v) \in R$, then $\alpha(u) = \alpha(v)$ and $(E_i(u),E_i(v)) \in R$ for every i $(1 \leq i \leq k)$.

If $E_i(u) \in V_I$ or $E_i(v) \in V_I$, we formally define that $(E_i(u),E_i(v)) \in R$ only in the case of $E_i(u) = E_i(v) \in V_I$, and $(E_i(u),E_i(v)) \notin R$ in other cases. If $E_i(u)$ or $E_i(v)$ are undefined, we formally define that $(E_i(u),E_i(v)) \in R$ only in the case that $E_i(u)$ and $E_i(v)$ are both undefined, and $(E_i(u),E_i(v)) \notin R$ in other cases.

S_A denotes the set of all SER's of $A = (G,\alpha)$. S_A is often denoted by S_G if A is element uniform.

Algorithm for construction of S_A

We assume here that V is finite and α is uniform. Because this Algorithm can be easily modified for the case where α is not uniform, we only show the Algorithm for uniform case. First, we define u_x for $u \in V$ and $x = E_{i_1}E_{i_2}\ldots E_{i_p} \in \{E_1,E_2,\ldots,E_k\}^*$ as follows: $u_x = E_{i_p}(\ldots(E_{i_2}(E_{i_1}(u)))\ldots)$.

For arbitrarily given SER S and vertex pair u and v such that $(u,v) \notin S$, we construct the minimum SER R such that $(u,v) \in R$ and $S < R$ (i.e. $(u,v) \in R$ and $(u,v) \notin R'$ for every refinement R' of R) by the following algorithm. We say that R is constructed from S and the vertex pair u and v.

Algorithm A

1) For every vertex pair p and q, let $(p,q) \in R$ if $(p,q) \in S$.

2) Let $(u,v) \in R$.

3) For all $x \in \{E_1,\ldots E_k\}^*$ such that $1 \leq |x| \leq |V|^2$ where $|V|$ is the number of vertices, let $(u_x,v_x) \in R$.

Proposition 3.1 For given SER S and vertex pair u and v such that $(u,v) \notin S$, Algorithm A gives the minimum SER R such that $(u,v) \in R$ and $S < R$.

Let $\mathcal{S}$ be a set of SER's. SER($\mathcal{S}$) denotes the set of all SER's each of which is constructed by the Algorithm A from an element R of $\mathcal{S}$ and a vertex pair u and v such that $(u,v) \notin R$.

Algorithm B

First we make the sequence of sets of SER's as follows:

$\mathcal{S}_0 = \{R_0\}, \mathcal{S}_1 = SER(\mathcal{S}_0), \ldots, \mathcal{S}_j = SER(\mathcal{S}_{j-1}), \ldots, \mathcal{S}_{m+1} = SER(\mathcal{S}_m) = \{R_U\}$ where $\mathcal{S}_m$ is the first set of SER's such that SER($\mathcal{S}_m$) is $\{R_U\}$. Let $S_A = \bigcup_{j=1}^{m+1} \mathcal{S}_j$.

Proposition 3.2 For a given automata network $A = (G,\alpha)$ where G is finite and α is uniform, Algorithm B gives S_A.

Next we consider the equivalence relation induced by the behavior of network.

Definition 3.3 State configurations which are consistent with an equivalence relation

A configuration $c \in C$ is defined to be consistent with an equivalence relation $R \in R_G$, iff if $(u,v) \in R$, then $c(u) = c(v)$.

Such a configuration is also called "R-consistent". C_R denotes the set of all R-consistent configurations.

Definition 3.4 Behavioral equivalence relation (BER)

For $A = (G, \alpha)$, $B \in R_G$ is said to be a behavioral equivalence relation (BER), if the following condition is satisfied.

If $(u,v) \in B$, then for every $c \in C_B$ and for every $\varphi \in \Phi$ $\hat{\alpha}_\varphi(c)(u) = \hat{\alpha}_\varphi(c)(v)$.

B_A is the set of all BER's of $A = (G, \alpha)$. we also write B_α in place of B_A if G is understood.

4. Structural Equivalence Relation in Cellular Automata

In this section, we show some results about S_G of the cellular automata. As we stated in section 2, a cellular automaton is an automata network which an identical automaton is placed at every vertex of a group graph.

Theorem 4.1 Let a graph $G = (V,N)$ be a group graph with $N = \{e_1, e_2, \ldots, e_k\}$ and H be a subgroup of V. If $V = \bigcup_p x_pH$ where x_pH is a coset of V, then the equivalence relation R induced by the partition $(x_1H, x_2H, \ldots)$ is in S_G.

Proof

For every vertex pair u and v, if $(u,v) \in R$ then there exists a coset x_pH such that $u = x_ph$, $v = x_ph'$ where $h,h' \in H$. So for every i $(1 \leq i \leq k)$, $(E_i(u), E_i(v)) = (e_iu, e_iv) = (e_i(x_ph), e_i(x_ph')) = ((e_ix_p)h, (e_ix_p)h') = (x_qh, x_qh')$ where $x_q = e_ix_p$. Then $E_i(u)$ and $E_i(v)$ are both in the coset x_qH. Therefore $(E_i(u), E_i(v)) \in R$. So $R \in S_G$.

Lemma 4.2 Suppose that a group graph $G = (V,N)$ is strongly connected and let R be an arbitrary equivalence relation from S_G. Then for every vertex pair u and v, if there exists an edge label i such that $(E_i(u), E_i(v)) \in R$ then $(u,v) \in R$.

Note In the following part of this paper, we write "concatenation of edges (of the path) from x to y" in place of "concatenation of neighborhood elements which correspond to edge labels (of the path) from x to y".

Corollary 4.3 If $(u,v) \in R$ then $(e_i^{-1}u, e_i^{-1}v) \in R$.

Corollary 4.4

Let $x = e_{i_1} \ldots e_{i_m}$. If $(u,v) \in R$ then $(xu, xv) \in R$ and $(x^{-1}u, x^{-1}v) \in R$.

Theorem 4.5 Let $G = (V,N)$ be a strongly connected group graph and the partition $(H_1, H_2, \ldots)$ induce the equivalence relation R in S_G, where $H_1 \ni \lambda$ (the identity of V). Then H_1 is a subgroup of V. Furthermore $H_i = x_iH_1$, then $V = \bigcup_i H_i = \bigcup_i x_iH_1$ where $x_i \in V$ and $x_1 = \lambda$.

Proof

1) H_1 is a subgroup of V.

First, for every vertex pair x, y $\in H_1$, we show xy $\in H_1$. Since the concatenation of edges from x to λ is x and $\lambda \in H_1$, if $(\lambda,y) \in R$ then $(x\lambda,xy) = (x,xy) \in R$. (Corollary 4.4) Therefore $xy \in H_1$. Second we show $x^{-1} \in H_1$ for every $x \in H_1$. Since the concatenation of edges from λ to x is x^{-1}, if $(\lambda,x) \in R$ then $(x^{-1}\lambda, x^{-1}x) = (x^{-1},\lambda) \in R$. So $x^{-1} \in H_1$

2) $H_i = x_iH_1$ and $V = \bigcup_i H_i = \bigcup_i x_iH_1$.

First we show that $H_i \supseteq x_iH_1$ where $x_i \in H_i$. Since x_i is the concatenation of edges from x_i to λ, and $(\lambda,u) \in R$ for every $u \in H_1$. Then $(x_i\lambda, x_iu) = (x_i, x_iu) \in R$. So $x_iu \in H_i$. Therefore $H_i \supseteq x_iH_1$.

Next we show $H_i \subseteq x_iH_1$. Suppose that there exists $v \in H_i$ such that $v \neq x_iu$ for every $u \in H_1$. Since v and $v' = x_iu$ ($u \in H_1$) are in H_i, $(x_iu,v) \in R$ then $(x_i^{-1}x_iu, x_i^{-1}v) = (u, x_i^{-1}v) \in R$. Then $x_i^{-1}v \in H_1$. So $v \in x_iH_1$. This is contradiction. So $H_i \subseteq x_iH_1$. Therefore $H_i = x_iH_1$. Then $V = \bigcup_i H_i = \bigcup_i x_iH_1$ is trivial. ∎

5. Relationships between S_A, B_A and $\mathbb{R}_G$

Proposition 5.1 For every $A = (G,\alpha)$, $S_A \subseteq B_A \subseteq \mathbb{R}_G$.

Proposition 5.2 If α is uniform, then for every G, $S_G = \bigcap_\alpha B_\alpha$.

Proposition 5.3 There is an automata network $A = (G,\alpha)$ such that $S_A \subsetneq B_A \subsetneq R_G$.

Definition 5.1 One-pair relation

A relation $R \in \mathbb{R}_G$ is said to be one-pair relation, if there exist $v_1, v_2 \in V$ such that $R = (\{v_1,v_2\},\{v_3\},\ldots,\{v_n\},\ldots)$ where $V = \{v_1,v_2, v_3,\ldots,v_n,\ldots\}$.

Let OP_{uv} denote a one-pair relation which satisfies that $u \neq v$, $(u,v) \in OP_{uv}$. By $\mathbb{R}_1$ we denote the set of all one pair relations, where the set of vertices is understood.

Proposition 5.4 If $\mathbb{R}_1 \subset S_A$ (or B_A), then S_A (or B_A) $= \mathbb{R}_G$.

In some cases, however, there exists a non trivial equivalence relation R (i.e. $R \neq R_0$, $R \neq R_U$) in S_A, while S_A contains no one-pair relation. In fact we have the following proposition.

Proposition 5.5 There exists an automata network $A = (G,\alpha)$ such that for any one-pair relation $OP_{uv} \notin S_A$ and $S_A \supsetneq \{R_0,R_U\}$.

Proposition 5.6 Let $A = (G,\alpha)$. $S_A = \mathbb{R}_G$ iff the following conditions holds: α is uniform and for every i ($1 \leq i \leq k$): $E_i(u) = u$ for every $u \in V$ or there exists a certain vertex $v \in V'$ such that $E_i(u) = v$ for every $u \in V$.

Proposition 5.7 If α is uniform and the element automaton is autonomous, then for every G $B_A = R_G$

Proposition 5.8 Suppose that α is uniform and not autonomous. If G is a graph such that indegree of every vertex is one, then $B_A = S_A$.

Proposition 5.9 There is an automata network $A = (G,\alpha)$ such that α is uniform, G is a graph where the indegree of every vertex is two, and $S_A = B_A \neq R_G$.

From the above Propositions we can consider the following Theorem. Since some preparations are necessary for it, we show it without proof.

Theorem 5.10 For every graph G such that $S_G \neq R_G$ and the indegree of every vertex of G is more than one, there exists a nonautonomous and uniform α , which maps every vertex to a "2 state automaton", such that $S_G \subsetneq B_\alpha$.

6. Concluding remarks

We showed some interesting results about the element uniform network. Though we have not yet obtained interesting result about the "nonuniform" network, both equivalence relations studied here are useful tools for studying automata networks.

From Proposition 5.6, we obtained the property of the graph such that $S_G = R_G$. We are also interested in the properties of the graphs such that $S_G = \{R_0,R_U\}$.

7. References

[1] Burks, A.W.: Essays on cellular automata. Illinois Press, 1970.

[2] Jump, J.R. and Kirtane, J.S.: On the Interconnection Structure of Cellular Networks. Information and Control 24, 74-91 (1974).

[3] Saito, T and Nishio, H.:LA Summer symposium report. in Japanese (1983).

[4] Saito, T and Nishio, H.:LA Winter symposium report. (1984).

[5] Saito, T and Nishio, H.:LA Summer symposium report. (1984).

ON A PUBLIC-KEY CRYPTOSYSTEM BASED ON PARALLEL REWRITING

Arto Salomaa[1)]

1. Introduction. A property shared by all classical cryptosystems is that, whenever you know the encryption key, e_K, then you also know the decryption key, d_K. More explicitly, d_K is either immediately implied by e_K or else can be computed from e_K without too much difficulty. This means that you must keep your encryption key e_K secret: once you publicize e_K, you give away the decryption key d_K.

The core idea behind public-key cryptosystems is that a cryptosystem may be constructed in such a way that the knowledge of e_K does not give away d_K. More specifically, the cryptanalyst may know e_K and hence, in principle, also d_K. However, the computation of d_K from e_K is intractable, at least for almost all keys K. Hence, for all practical purposes, the encryption of the plaintext remains secure against cryptanalytic attacks, although the encryption key e_K is made public.

The reader is referred to [6] for a discussion of the various advantages inherent in the usage of public-key cryptosystems.

After the publication of the seminal paper [2] by Diffie and Hellman, the two public-key systems most widely studied have been the RSA system, [4], and the knapsack system, [3]. (See also [9] which is the most important cryptanalytic paper concerning the knapsack system.) Although the idea of a "one-way" function e_K is by no means restricted to number theory, both of the public-key systems mentioned are based on number-theoretic concepts.

However, concepts from other fields have also been used in the construction of public-key cryptosystems. A system based on ideas from formal language theory (more specifically, parallel rewriting in the sense of L systems) was presented in [8]. The reader is referred also to [6] or [7] for more information concerning this system. A special advantage of the system is that it is very unlikely, if not impossible, to find the secret decryption key from the public encryption key by preprocessing only, i.e., without knowing samples of the ciphertext. The reason for this is that there is no way of recognizing a proposed decryption key to be the true one. For instance, in the RSA system you immediately know when you succeeded in your cryptanalysis: it means the factorization of the product of two large primes.

The purpose of this paper is a further discussion of the system presented in [8]. We have tried to make the paper as self-contained as possible. After some general remarks about the construction of public-key cryptosystems in Section 2, the system of [8] is described together with

1) Mathematics Department, University of Turku, Finland

an illustrative example in Section 3. Section 4 contains discussion, scattered remarks and open problems, mainly from a cryptanalytic point of view.

2. The construction of public-key cryptosystems: an outline. The construction is based on the idea of a trapdoor: the information publicized is not enough without the knowledge of a secret trapdoor.

More specifically, the trapdoor technique works as follows in the construction of public-key cryptosystems.

(i) Start with a difficult (intractable) problem P.

(ii) Consider an easy subproblem P_1 of P. P_1 should be in polynomial time, preferably in linear time.

(iii) "Shuffle" P_1 in such a way that the resulting problem $\overline{P_1}$ "looks like" the original problem P. Use $\overline{P_1}$ as a public encryption key.

(iv) Keep secret the information concerning how P_1 can be recovered from $\overline{P_1}$. This information is referred to as the trapdoor.

Of course, this technique is still on the abstract level: in practice the details have to be filled in. However, the so far existing public key cryptosystems have been constructed by this technique. Moreover, in so far existing examples the initial example P is at most *NP*-complete. It is conceivable, and even desirable, that the initial problem lies higher up in the complexity hierarchy. It could be even an undecidable problem. Observe that $\overline{P_1}$ only "looks like" P but can never be the same problem.

Typical candidates for problems P and P_1 would be the (undecidable) 0L equivalence problem and the (easy from the point of view of cryptography) D0L equivalence problem. However, in spite of many efforts, no feasible cryptosystem has been constructed on this basis.

The cryptosystem described in the next section is based on a similar difference between arbitrary and deterministic L systems: the membership problem for T0L and DT0L systems, the former being *NP*-complete and, the latter, polynomial time.

The reader is referred to [5] for all unexplained notions concerning L systems. However, this background is not really necessary for understanding the basics of this paper because we try to describe everything in quite general terms.

3. The cryptosystem and an example. Let Σ be an alphabet and consider two morphisms $h_0, h_1 : \Sigma^* \rightarrow \Sigma^*$, as well as a nonempty word w in Σ^*. For a word $i_0 i_1 \dots i_n$ over the alphabet $\{0,1\}$, where each i is a letter, and a word x over Σ we denote briefly

$$(x)i_0 i_1 \dots i_n = h_{i_n}(\dots h_{i_1}(h_{i_0}(x)) \dots) .$$

We assume that our morphisms are unambiguous in the following sense. Whenever

$$(w)i_0 i_1 \dots i_n = (w)j_0 j_1 \dots j_m$$

then

$$i_0 i_1 \dots i_n = j_0 j_1 \dots j_m .$$

Let now Δ be an alphabet of a much greater cardinality than Σ. (Typically, Σ could consist of two letters whereas Δ consists of 200 letters.) Let $g : \Delta^* \to \Sigma^*$ be a morphism mapping every letter to a letter or to the empty word in such a way that $g^{-1}(a)$ is nonempty for all letters a in Σ. Define two substitutions σ_0 and σ_1 on Δ^* by letting $\sigma_i(d)$ to be a finite nonempty subset of $g^{-1}(h_i(g(d)))$, for all d in Δ and $i = 0, 1$.

Intuitively, the morphism g divides the letters of Δ into "descendants" of letters of Σ and into "junk" symbols that are mapped into the empty word.

Finally, choose a word u from $g^{-1}(w)$. The quadruple $(\Delta, \sigma_0, \sigma_1, u)$ now constitutes the <u>public encryption key</u>. The encryption of a plaintext $i_0 i_1 \dots i_n$ (in bits) is carried out by choosing an arbitrary word from

$$(*) \qquad \sigma_{i_n}(\dots \sigma_{i_1}(\sigma_{i_0}(u)) \dots) .$$

The plaintext can also be divided into blocks (of not necessarily equal length) that will be encrypted separately by the method described. This latter procedure is recommended if the cryptotext tends to be too long.

The <u>secret decryption key</u> consists of the items Σ, h_0, h_1, w and g. In fact, g is the essential item because the other items can be determined from g and the public encryption key.

It follows from the definition of the public encryption key that, whenever x is a word in the set (*), then

$$g(x) = (w)i_0 i_1 \dots i_n .$$

The original plaintext can now be recovered from g(x), using the items Σ, h_0, h_1 and w. (In the terminology of L systems, we make use of the DTOL system determined by these four items.)

We will now consider a specific example. The morphisms and substitutions are defined in terms of productions. This notation is used to aid intuition.

Assume that $\Sigma = \{a,b\}$, and that h_0 and h_1 are defined by

$$h_0 : a \to ab ,\ b \to b \quad ; \quad h_1 : a \to a ,\ b \to ba .$$

Choose $w = ab$.

Consider $\Delta = \{c_1, c_2, c_3, c_4, c_5\}$. The alphabet Δ is much too small for any practical purposes. However, practically feasible alphabets would result in systems unreadable in an article.

Define

$$g(c_2) = g(c_4) = a, \quad g(c_1) = b, \quad g(c_3) = g(c_5) = \lambda .$$

Thus, c_2 and c_4 are descendants of a, c_1 is the only descendant of b, whereas c_3 and c_5 are junk letters.

Define σ_0 and σ_1 by

$$\sigma_0 : c_1 \to c_5c_1, \; c_2 \to c_2c_3c_1, \; c_2 \to c_4c_5c_1, \; c_3 \to c_3c_5,$$
$$c_3 \to c_5c_3, \; c_4 \to c_2c_1c_3c_5, \; c_5 \to c_5c_5c_5 ;$$

$$\sigma_1 : c_1 \to c_1c_2, \; c_1 \to c_1c_4, \; c_2 \to c_4, \; c_3 \to c_3c_5c_3,$$
$$c_3 \to c_5c_5, \; c_4 \to c_2, \; c_4 \to c_4, \; c_5 \to c_5 .$$

Finally, define $u = c_2c_3c_1$. The quadruple $(\Delta, \sigma_0, \sigma_1, u)$ can now be publicized as the encryption key.

Consider the plaintext 010. Thus, the encryption must start from u and use the substitutions $\sigma_0, \sigma_1, \sigma_0$. One possible "derivation" goes as follows:

$$u = c_2c_3c_1 \underset{\sigma_0}{\Rightarrow} c_2c_3c_1c_3c_5c_5c_1$$
$$\underset{\sigma_1}{\Rightarrow} c_4c_5c_5c_1c_2c_3c_5c_3c_5c_5c_1c_2$$
$$\underset{\sigma_0}{\Rightarrow} c_2c_1c_3c_5c_5^3c_5^3c_5c_1c_2c_3c_1c_3c_5c_5^3c_5c_3c_5^3c_5^3c_5c_1c_4c_5c_1$$
$$= c_2c_1c_3c_5^8c_1c_2c_3c_1c_3c_5^5c_3c_5^7c_1c_4c_5c_1 = x .$$

Thus, the word x constitutes an encryption of the plaintext 010. Clearly, x is not the only possible encryption since there are some alternatives when applying the substitutions σ_i. For instance, the second word might be $c_4c_5c_1c_5c_3c_5c_1$.

Let us now consider decryption based on the trapdoor. We, thus, know a word y over the alphabet $\Sigma = \{a,b\}$, where y is obtained by applying the morphism g to the cryptotext. We also know that y is of the form

$$(5) \qquad y = (w)i_0i_1 \ldots i_n .$$

On the basis of this knowledge, we have to recover the plaintext $i_0i_1 \ldots i_n$.

Observe first that, whenever we know h_i ($i = 0$ or $i = 1$) and $w' = h_i(w'')$, then w'' is unique and can be immediately computed. This follows by the definition of h_i: "parsing" is unique and can be done fast. On the other hand, the value of $i = 0,1$ can be immediately found from w' . Continuing in this way, decryption can be accomplished in linear time.

We mention, finally, that it is not necessary to begin with two morphisms h_0 and h_1 . For instance, if one wants to use decimal notation in the plaintext then it is natural to begin with ten morphisms and construct ten substitutions as well.

4. Discussion and open problems. An advantage of the cryptosystem discussed in the previous section is that it is very unlikely to find the secret decryption key directly from the public encryption key. This is due to the fact that there is no way of recognizing a candidate for the decryption key to be the correct decryption key. A further discussion of this aspect now follows.

An obvious cryptanalytic approach is the following. Junk letters can produce only junk letters (whereas non-junk letters can produce both types of letters). Thus, by forming "transitive closures" one can obtain all possible candidates for the set of junk letters. After this, it is easier to obtain the underlying morphisms, since only real descendants have to be taken into account.

Assume now that the cryptanalyst is able to find a DTOL system G' such that the publicized TOL system G results from G' according to the rules given in Section 3. The cryptanalyst does not, of course, know whether G' is the initial DTOL system G" used by the cryptosystem designer. However, the following result can be obtained.

Assume that G' and G" are as above. If G' is "backward deterministic" (unambiguous) in the sense of Section 3 then it yields the same decryption results as G'. On the other hand, if G' is not backward deterministic, then it is possible that the correct decryption result (obtainable according to G") is not among all the decryption results obtainable according to G'.

Recall that G" was assumed to be backward deterministic. This means that, whenever G" is used to decrypt a ciphertext, at most one plaintext results. The result above shows that if also G' possesses this property of backward determinism, then it can be used for decryption, although the cryptosystem designer has used a different system G". (Compare this state of affairs with the discussion given in [9] about different trapdoor pairs.)

But how does our cryptanalyst know whether or not G' is backward deterministic? In general, this problem has been shown undecidable in [1]. Moreover, if G' is not backward deterministic, then decrypting according to G' might be almost equally difficult as decrypting according to G. So nothing much is gained although such a G' is found, in particular, because all decryption results may be wrong.

It remains an open problem, basically in the theory of DTOL systems, to exhibit a reasonably big class of systems satisfying the requirements: the systems should be backward deterministic, as well as easily parsable and should also not increase the word length too rapidly. The last requirement is added to prevent the ciphertext becoming too long in comparison with the plaintext.

Acknowledgement. The author is grateful to Valtteri Niemi for useful discussions.

References

[1] J. DASSOW. A note on DTOL systems. Manuscript (1983).

[2] W. DIFFIE and M. HELLMAN. New directions in cryptography. IEEEE Translations on Information Theory 22 (1976) 644-654.

[3] R. MERKLE and M. HELLMAN. Hiding information and signatures in trapdoor knapsacks. Ibid. 24 (1978) 525-530.

[4] R. RIVEST, A. SHAMIR and L. ADLEMAN. A method for obtaining digital signatures and public-key cryptosystems. ACM Communications 21 (1978) 120-126.

[5] G. ROZENBERG and A. SALOMAA. The Mathematical Theory of L Systems. Academic Press (1980).

[6] A. SALOMAA. Computation and Automata. Encyclopedia of Mathematics, Cambridge University Press (in press).

[7] A. SALOMAA. Trapdoors and protocols: recent trends in cryptography. In: H. Maurer (ed.) "Überblicke Informationsverarbeitung 1984", BI Mannheim-Wien-Zürich (1984) 275-320.

[8] A. SALOMAA and E. WELZL. A cryptographic trapdoor based on iterated morphisms. Manuscript (1983).

[9] A. SHAMIR. A polynomial time algorithm for breaking Merkle-Hellman cryptosystems. Proceedings of the 23rd FOCS Symposium (1982) 145-152.

PARALLEL METHODS FOR NONLINEAR OPTIMISATION

U. Schendel, M. Schyska [1)]

During the past ten years increased efforts have been made to construct parallel algorithms for nonlinear optimisation. The procedures described here are due to Sloboda [9] and Chazan and Miranker [2] , and do not require the derivatives of the function to be minimised. They are based on a serial conjugate Gram-Schmidt procedure. The underlying notion is to calculate the function values in parallel in order to reduce the time needed by the minimisation procedure. The following summary is based on studies by Schendel and Schyska [7] , [8] .

INTRODUCTION

Most serial procedures for the minimisation of a function f of n real variables generate iteratively a sequence of n-vectors x_k (k=2,3,...) of real components starting from a suitable initial vector x_1. Each step of the iteration consists of two parts:

Step 1: Calculate according to prescribed rules a search direction p_k, a real n-vector, where the subscript denotes the iteration currently in progress.

Step 2: Determine by means of a search along a line (linear search) a step length $a_k \in \mathbb{R}$ such that, at least approximately,

$$f(x_k+a_kp_k) = \min_a f(x_k+ap_k) ,$$

and then set

$$x_{k+1} = x_k + a_kp_k \qquad \square$$

Individual procedures differ essentially only in the way the search direction is chosen. Conjugate direction procedures are such that for a quadratic function the global minimum is reached using exact linear search in not more than n steps, where n is the number of variables in the problem.

CONJUGATE GRAM-SCHMIDT-PROCEDURE

For the purposes of this introductory description of the Gram-Schmidt procedure it will be assumed that the function to be minimised $f:\mathbb{R}^n \to \mathbb{R}$ is the quadratic form

$$f(x) = 1/2\, x^TAx - b^Tx+c , \qquad (1)$$

---

1) Institut für Mathematik III, Freie Universität Berlin, D-1000 Berlin

where A is a symmetric n×n matrix of real elements, b is a real n-vector, c is a real number. The unique global minimum point (real n-vector) x^* satisfies

$$x^* = A^{-1}b \ . \tag{2}$$

To save repetition it may be assumed that all matrices and vectors and scalars have real components, unless otherwise stated. Moreover it will be assumed that an n-dimensional problem is being handled.

An important property of quadratic functions is the uniqueness of minimum points on lines. In this connection we enunciate the following

<u>Lemma:</u>
The minimum point x_2 of (1) on the line

$$L := \{x \mid x = x_1 + \alpha p_1, \alpha \in \mathbb{R}\}$$

passing through x_1 in the direction p_1 is given uniquely by

$$x_2 = x_1 + a_1 p_1$$

where the step length a_1 is given by

$$a_1 = -\frac{p_1^T(Ax_1-b)}{p_1^T A p_1} \ . \qquad \square$$

Next we recall the concept of pairwise conjugacy of vectors with respect to a matrix. This is a generalisation of orthogonality. Thus, the orthogonalisation of the Gram-Schmidt procedure can be adapted to the construction of n pairwise conjugate vectors $\hat{p}_1, \ldots, \hat{p}_n$ with respect to the matrix A from n linearly independent vectors $u_1, \ldots, u_n$. Using the rule

$$\begin{aligned} \hat{p}_1 &= u_1 \ , \\ \hat{p}_2 &= u_2 - b_{21}\hat{p}_1 \ , \\ &\vdots \\ \hat{p}_n &= u_n - b_{n1}\hat{p}_1 - b_{n2}\hat{p}_2 - \ldots - b_{n,n-1}\hat{p}_{n-1} \ , \end{aligned}$$

we have, from the conjugacy requirement

$$\hat{p}_i^T A \hat{p}_k = 0 \qquad (k = 2,\ldots,n; \ i = 1,2,\ldots,k-1),$$

the result

$$b_{ki} = \frac{u_k^T A \hat{p}_i}{\hat{p}_i^T A \hat{p}_i} \qquad (k = 2,\ldots,n; \ i = 1,\ldots,k-1)$$

for the conjugacy coefficients.

This leads to

<u>Algorithm 1</u>
An initial value (point) (3)

$$x_1 \in \mathbb{R}^n$$

is given, together with n linearly independent vectors

$$u_1, u_2, \ldots, u_n \in \mathbb{R}^n \tag{4}$$

The kth step of the iteration is then broken down into:

<u>Step 1:</u> Calculate the kth search direction $\hat{p}_k$ from

$$\hat{p}_k = u_k - \sum_{i=1}^{k-1} b_{ki}\hat{p}_i \tag{5}$$

with

$$b_{ki} = \frac{u_k^T A \hat{p}_i}{\hat{p}_i^T A \hat{p}_i}, \quad i = 1,\ldots,k-1 . \tag{6}$$

<u>Step 2:</u> Determine by one line search starting from the point x_k the step length

$$a_k \in \mathbb{R} \tag{7}$$

and put

$$x_{k+1} = x_k + a_k\hat{p}_k \tag{8}$$ □

With exact line search the algorithm terminates at the global minimum point of (1), that is

$$x_{n+1} = x^* , \tag{9}$$

from the pairwise conjugacy of the search directions $\hat{p}_1,\ldots,\hat{p}_n$ with respect to the matrix A (see, for example, [ ]).

Formulae (5) for the computation of the search directions form a linear recurrence R<n,n-1> of order n-1 with n variables. Several parallel algorithms have been developed for the solution of such recurrences. Particularly suitable for R<n,n-1> is the Column-Sweep Algorithm and it can thus be used to compute the search directions $\hat{p}_1,\ldots,\hat{p}_n$.

Figure 1 shows the procedure diagrammatically.

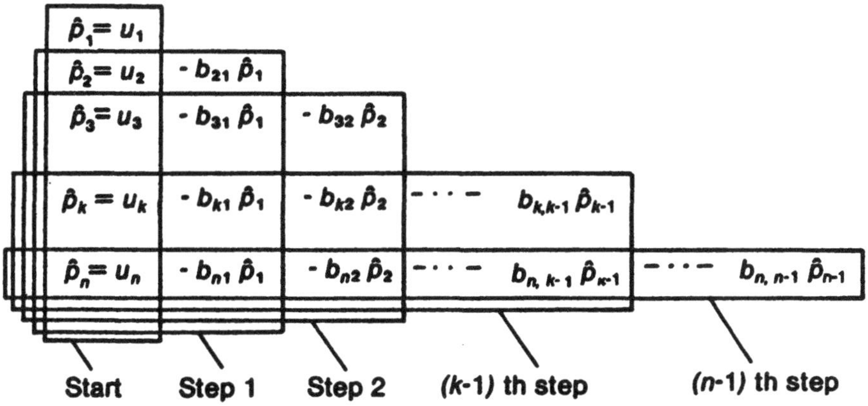

Fig. 1 -Illustrating the procedure for the parallel computation of(5).

To begin, the following assignment is made in parallel for $i=1,\ldots,n$:

$$\hat{p}_i^{(o)} = u_i \quad . \tag{10}$$

Then the coefficients

$$b_{ik} = \frac{u_i^T A \hat{p}_k}{\hat{p}_k^T A \hat{p}_k} = \frac{\hat{p}_i^{(k-1)^T} A \hat{p}_k^{(k-1)}}{\hat{p}_k^{(k-1)^T} A \hat{p}_k^{(k-1)}} \tag{11}$$

are computed one after the other for $k = 1,\ldots,(n-1)$, and at each step this is done in parallel for $i = k+1,\ldots,n$. Following this, the vectors

$$\hat{p}_i^{(k)} = \hat{p}_i^{(k-1)} - b_{ik}\hat{p}_k^{(k-1)} \tag{12}$$

are calculated, again in parallel, for $i = k+1,\ldots,n$. From (5) and (6) it follows that for all $k = 1,\ldots,n$

$$\hat{p}_k^{(k-1)} = \hat{p}_k \quad . \tag{13}$$

The following Lemma from [4] explains how conjugate directions can be calculated without explicit knowledge of the derivatives of the function to be minimised.

Lemma:

Let f be a quadratic function of form (1) and let $x_2, \bar{x}_2$ be minimum points of f on the lines

$$L = \{x \mid x=x_1+\alpha p_1\} \tag{14}$$

$$L = \{x \mid x=\bar{x}_1+\alpha p_1\} \tag{15}$$

through x_1 and $\bar{x}_1$ respectively with the same direction vector p_1. Then

$$(x_2-\bar{x}_2)^T A\, p_1 = 0 \tag{16}$$ □

THE ALGORITHM OF SLOBODA

By suitable choice of starting points for line searches the search directions of the conjugate Gram-Schmidt procedure can also be calculated in terms of differences of minimum points of f on lines instead of using the formulae (5), (6), (11), (12). If, moreover, the line searches for the calculation of these minimum points are carried out in parallel we obtain the parallel procedure of Sloboda which will be described next. In order to compare the two processes the function f will continue to be of form (1).

Algorithm 2

Given an arbitrary starting point

$$x_1^{(o)} \in \mathbb{R}^n \tag{17}$$

and n linearly independent vectors

$$u_1,\ldots,u_n \in \mathbb{R}^n \tag{18}$$

Start: In parallel set

$$x_1^{(i)} = x_1^{(o)} + u_i \qquad (i = 1,\dots,n). \tag{19}$$

At iteration step k (k = 1,...,n):

Step 1: Put

$$p_k = x_k^{(k)} - x_k^{(k-1)} . \tag{20}$$

Step 2: Determine by n-k+1 line searches implemented in parallel starting from $x_k^{(k)},\dots,x_k^{(n)}$ in the common direction p_k the n-k+1 step lengths

$$a_k^{(i)} \in \mathbb{R} \qquad (i = k,k+1,\dots,n) \tag{21}$$

and in parallel for $i = k,k+1,\dots,n$ assign

$$x_{k+1}^{(i)} = x_k^{(i)} + a_k^{(i)} p_k . \tag{22}$$ □

The equivalence of the search directions of the procedures of Sloboda and Gram-Schmidt follows using (12) from

$$x_k^{(i)} - x_k^{(k-1)} = \hat{\beta}_i^{(k-1)} \qquad k = 1,\dots,n;\ i = k,\dots,n. \tag{23}$$

Because of the uniqueness of the minimum points of quadratic functions on lines and the procedure used in (20) to obtain the search directions p_k, Step 2 can be substituted by

Step 2a: Determine by n-k+1 line searches implemented in parallel starting from the points $x_k^{(k-1)},\ x_k^{(k+1)},\ x_k^{(k+2)},\dots,x_k^{(n)}$ and all in the common direction p_k, the n-k+1 step lengths

$$a_k^{(k-1)},\ a_k^{(k+1)},\ a_k^{(k+2)},\dots,a_k^{(n)} \in \mathbb{R} \tag{24}$$

and in parallel assign

$$x_{k+1}^{(k)} = x_k^{(k-1)} + a_k^{(k-1)} p_k \tag{25}$$

and for $i = k+1,\dots,n$,

$$x_{k+1}^{i} = x_k^{(i)} + a_k^{(i)} p_k . \tag{26}$$ □

The connection between the iteration points and search directions is clarified by Fig. 2 .

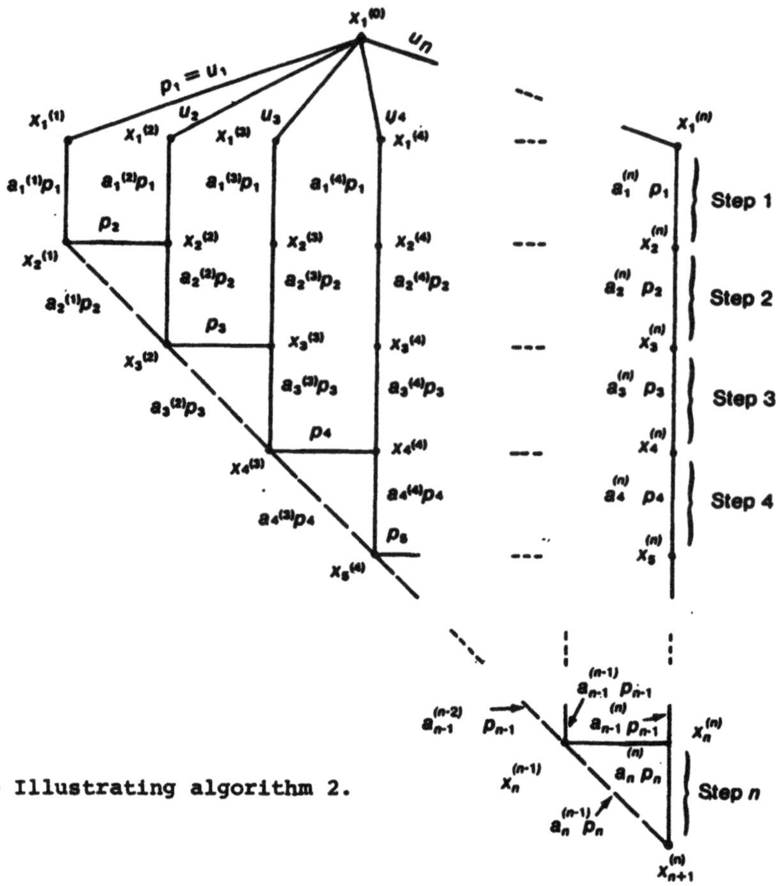

Fig. 2 - Illustrating algorithm 2.

By (22) and (25) the following hold for the final iteration point:

$$x_{n+1}^{(n)} = x_1^{(n)} + \sum_{i=1}^{n} a_i^{(n)} p_i$$
$$= x_1^{(o)} + \sum_{i=1}^{n} a_i^{(i-1)} p_i \ , \qquad (27)$$

that is, $x_{n+1}^{(n)}$ is obtained by a sequence of n line searches in the directions $p_1,\dots,p_n$ beginning at the initial point $x_1^{(n)}$, or the equivalent with $x_1^{(o)}$. Because of the conjugacy property of the search directions the global minimum point x^* of f is obtained at $x_{n+1}^{(n)}$ with exact line search. From the equality of the search directions it also follows that Sloboda's procedure, starting from the same point as the conjugate Gram-Schmidt procedure, computes the same sequence of minimum points $\{x_1^{(o)}, x_2^{(1)}, \dots, x_{n+1}^{(n)}\}$ as the sequence $\{x_1, x_2, \dots, x_{n+1}\}$ computed by conjugate Gram-Schmidt.

For the minimisation of a function f which is not quadratic by the conjujate Gram-Schmidt, or Sloboda, procedures the cycles described in Algorithms 1 and 2 of n iteration steps are repeated until a sufficiently close approximation to a global minimum point of f is achieved. The last iteration point x_{n+1}, or $x_{n+1}^{(n)}$ calculated in a cycle then serves as the starting point for the next cycle. The conjugate Gram-Schmidt and Sloboda procedures are thus restart procedures.

THE ALGORITHM OF CHAZAN AND MIRANKER

A parallel minimisation procedure which is not a restart procedure is the procedure of Chazan and Miranker. When f is quadratic this constructs the same search directions as the conjugate Gram-Schmidt.

Algorithm 3

An arbitrary starting point

$$x_1^{(o)} \in \mathbb{R}^n \tag{28}$$

is given together with n linearly independent vectors

$$u_1, u_2, \ldots, u_n \in \mathbb{R}^n . \tag{29}$$

Start: For $i = 1, \ldots, n$ set

$$x_1^{(i)} = x_1^{(i-1)} + u_i . \tag{30}$$

At iteration step k $(k = 1, 2, \ldots)$:

Step 1: Put

$$p_k = x_k^{(1)} - x_k^{(o)} . \tag{31}$$

Step 2: Determine by n line searches implemented in parallel starting from $x_k^{(1)}, \ldots, x_k^{(n)}$ in the common direction p_k the n step lengths

$$a_k^{(i)} \in \mathbb{R} \qquad (i = 1, \ldots, n) \tag{32}$$

and in parallel for $i = 1, \ldots, n$ assign

$$x_{k+1}^{(i-1} = x_k^{(i)} + a_k^{(i)} p_k . \tag{33}$$

Step 3: Put

$$x_{k+1}^{(n)} = x_{k+1}^{(n-1)} + u_j \tag{34}$$

where

$$j = ((k-1) \bmod n) + 1 . \tag{35}$$ □

It can be shown (see [ ]) that when f is quadratic exact line search with (12) makes the following hold:

$$x_k^{(i)} - x_k^{(i-1)} = \hat{\beta}_{(k-1+i)}^{(k-1)}, \quad k = 1, \ldots, n; i = 1, \ldots, n-k+1$$

and in the special case of (13) (36)

$$p_k = x_k^{(1)} - x_k^{(o)} = \hat{\beta}_k^{(k-1)} = \hat{\beta}_k, \quad k = 1, \ldots, n . \tag{37}$$

Because of the construction used Step 2 can be replaced by:

Step 2a: Determine by n line searches implemented in parallel starting from $x_k^{(o)}, x_k^{(2)}, x_k^{(3)}, \ldots, x_k^{(n)}$ in the common direction p_k the n step lengths

$$a_k^{(o)}, a_k^{(2)}, \ldots, a_k^{(n)} \in \mathbb{R} \tag{38}$$

and in parallel assign

$$x_{k+1}^{(o)} = x_k^{(o)} + a_k^{(o)} p_k \quad , \tag{39}$$

and for $i = 2, \ldots, n$

$$x_{k+1}^{(i-1)} = x_k^{(i)} + a_k^{(i)} p_k \ . \tag{40}$$ □

Fig. 3 explains this.

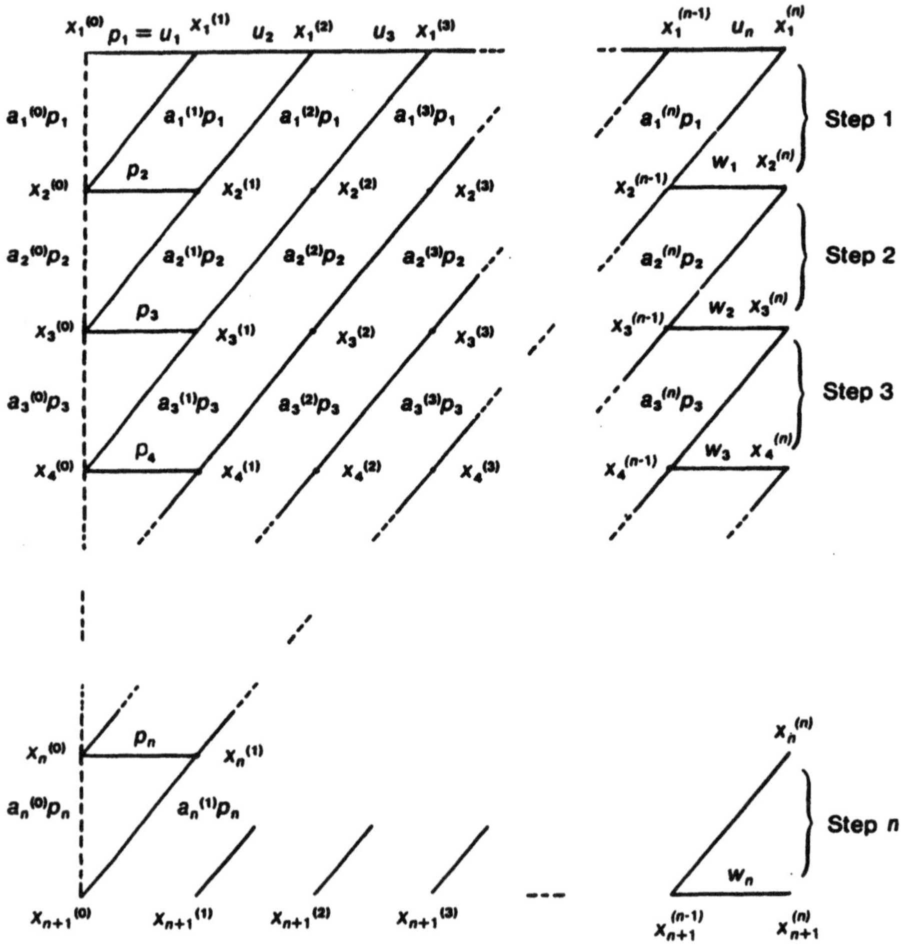

Fig. 3 -Illustrating algorithm 3 .

By (33) and (39) we have

$$x_{n+1}^{(o)} = x_1^{(n)} + \sum_{i=1}^{n} a_i^{(n-i+1)} p_i \tag{41}$$

$$= x_1^{(o)} + \sum_{i=1}^{n} a_i^{(o)} p_i \ .$$

By the conjugacy of the search directions $p_1,\ldots,p_n$ the global minimum point x^* of the quadratic function (1) is thus reached at $x_{n+1}^{(o)}$. Moreover starting with a common initial point and using exact line search, Algorithm 3 generates the sequence $\{x_1^{(o)},x_2^{(o)},\ldots,x_{n+1}^{(o)}\}$ of minimum points and this is the same as the sequence $\{x_1,x_2,\ldots,x_{n+1}\}$ given by the conjugate Gram-Schmidt procedure, and the sequence $\{x_1^{(o)},x_2^{(1)},\ldots,x_{n+1}^{(n)}\}$ given by the Sloboda procedure.

NUMERICAL RESULTS

The described algorithms have been tested with the following test functions:

Function 1: $f : \mathbb{R}^2 \rightarrow \mathbb{R}$

$$f(x) = 100(x_1^2 - x_2)^2 + (1-x_1)^2$$

Function 2: $f : \mathbb{R}^{20} \rightarrow \mathbb{R}$

$$f(x) = \sum_{i=1}^{10} 100(x_{2i} - x_{2i-1}^2)^2 + (1-x_{2i-1})^2$$

Parallel execution of the parallel algorithms 2 and 3 was simulated on a serial computer . As an estimation of the required execution time the number of function evaluations was used. To this end the total of function evaluations was counted in each of the simulated parallel working processors and the execution time of the parallel algorithms is determined by the maximum of all these amounts (f_{max}). Amongst this "parallel execution time" Table 1 shows the number of necessary iteration steps (I) and the total of all function evaluations (Σf) of the tested algorithms.

		I	f	f_{max}
function 1	CGS	30	1476	1476
	Sloboda	30	2122	1415
	Chazan/Miranker	30	1901	951
function 2	CGS	300	25 560	25 560
	Sloboda	280	138 880	13 228
	Chazan/Miranker	226	216 303	10 834

Table 1

The low speed-up results from the fact, that the number of that line searches of the parallel algorithms, which are performed one after the other, is approximately the same as that of the serial conjugate Gram-Schmidt algorithm, and these line searches are the most time consuming parts of the algorithms.

REFERENCES

[1] Brandenburger, J., Rekurrente Relationen in der Parallelen Numerik, Diplom-Thesis, Freie Universität Berlin (1983)

[2] Chazan, D. & Miranker, W.L., A Nongradient and Parallel Algorithm for Unconstrained Minimization, SIAM J. of Control, 8, 207-217 (1970)

[3] Feilmeier, M., Joubert, G. & Schendel, U. (eds.), Parallel Computing '83, Proceedings of the International Conference on Parallel Computing, held at the Freie Universität Berlin, 26-29 September 1983, North-Holland, Amsterdam-New York-Oxford (1984)

[4] Hestenes, M., Conjugate Direction Methods in Optimization, Applications of Mathematics, No. 12, Springer, New York-Heidelberg-Berlin (1980)

[5] Hockney, R.W. & Jesshope, C.R., Parallel Computers, Adam Hilger Limited, Bristol (1981)

[6] Schendel, U., Introduction to Numerical Methods for Parallel Computers, Ellis Horwood Limited, Chichester (1984)

[7] Schendel, U. & Schyska, M., Parallele Algorithmen in der Nichtlinearen Optimierung, Preprint No. 161/1984, Freie Universität Berlin (1984)

[8] Schyska, M., Parallele Algorithmen in der Nichtlinearen Optimierung, Diplom-Thesis, Freie Universität Berlin (1983)

[9] Sloboda, F., Parallel Method of Conjugate Directions in Minimization, Aplikace Matematiky, 20, 436-446 (1975)

DESCRIPTION, FUNCTIONAL ANALYSIS AND SIMULATION OF PARALLEL SYSTOLIC ARCHITECTURES

N. PETKOV TURKEDJIEV[1)]

INTRODUCTION

The advances in solid state technology have made it possible that hundreds of thousands of devices can be put on a single chip and that number will increase to several millions before the ultimate physical limits of the integration are reached. VLSI creates the opportunity to implement whole complex systems on single chips and in that way the creation of parallel computing systems, which are economically unjustified, if made with standard IC, becomes possible. Integrated circuit technology is most suitable for systems consisting of similar repeating elements and it seems to be the natural medium for adequate parallel architectures. In addition to improved cost, integration, power dissipation, and reliability, such single-chip parallel systems radically improve the performance by using a number of processing elements and have considerably greater computational power in comparison with the traditional von Neumann architectures.

It is generally accepted that VLSI complexity can be governed only by powerful design automation tools. However, most of these tools are very expensive to develop and support. That is why it is reasonable to determine and create design automation tools, which correspond to the problems arising in the implementation of parallel architectures. Here we consider systolic parallel architectures /1/, which are especially suitable for IC implementation. These systems consist of identical processing elements (cells), which are connected with regular local interconnections. These cellularity and regularity are the unique features of the systolic architectures, which determine the unique problems in their design and the corresponding design automation tools. In the following the design levels crucial for the design of IC implementing parallel systolic algorithms are considered and the adequate design automation tools are determined.

CRUCIAL DESIGN LEVELS

The traditional design problems at the lower design levels as logical, electrical, and physical (layout) design, are sugnificantly reduced in the case of systolic architectures, because only one processing element

---

1) Technische Universität Dresden, Sektion Informationstechnik, 8027 Dresden, Mommsenstraße 13, DDR

(cell) has to be designed. After that this cell is repeated as many times as necessary to form a regular pattern and the cells are interconnected. The relative positions of the cells and the length of the (local) interconnection paths are known in advance, so that repeatative redesigns of the cell are not necessary. If the cell is simple enough, its design and the design of the whole systolic system can be carried out with simple computational aids. In that way, the availability and the use of sophisticated design automation tools for logical, electrical and physical design and verification are not crucial for the implementation of systolic algorithms in silicon.

The advantages mentioned above feature the systolic algorithms as very attractive for IC implementation. Unfortunately, there is no general approach to the design of systolic algorithms. Untill now the design of such algorithms has been much more a matter of intuition than a matter of methodology. Some typical attributes such as cellular, regular structure, regular pattern of local interconnections, interaction of several data flows have emerged, but they are insufficient to make a methodology for designing systolic algorithms. Perhaps the lack of understanding how to design such systems is due to the difference between their parallel nature and the sequential nature of the human thinking, to the fact that we are not able to keep in mind all the changes taking place in a parallel processing system.

Let us after Kung /1/ consider a representative example - the hexagonal processor array for matrix multiplication (fig. 1). It consists of hexagonally connected equal processing elements performing the operation

$$ZC \leftarrow ZC + ZA.ZB \qquad (1)$$

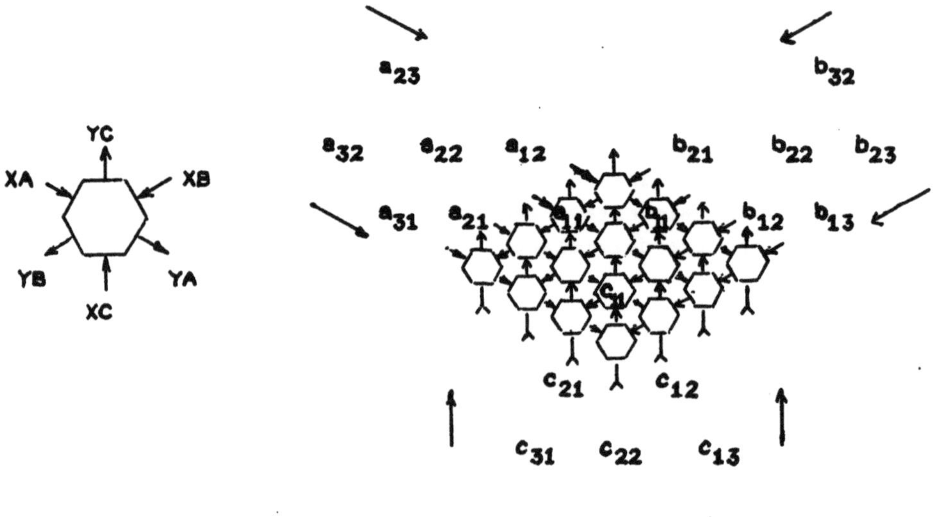

Fig. 1 - Processor array for matrix multiplication

Upon every new clock period data is input from the inputs XA, XB, and XC in the inner registers ZA, ZB, and ZC. The operation (1) is carried out and at the end of the clock period the data is transfered from the inner registers to the corresponding outputs YA, YB, and YC. If data is entered in a special way (shown on fig. 1), the array is capable of performing matrix multiplication of two band matrixes and the elements of the resulting matrix product appear at certain clock periods at the output ports of the array.

It is not obvious that the array should perform exactly that function and it is also not obvious when and where one must input data and when and where the output data is available. Anyway, it is obvious that one can not keep in mind all the changes that are taking place in the system. If we want to keep account of them and understand the work of the system, we shall have to carry out several consequtive steps, i. e. simulate the work of the system. That task seems to be a perfect job for a computer. That is why, it is reasonable to create computational aids, which would help us understand the behaviour of the systolic systems and support the development of new systolic algorithms and architectures.

FUNCTIONAL ANALYSIS AND SIMULATION OF SYSTOLIC ARCHITECTURES

A number of design automation tools have been developed to support the functional simulation of computer systems /2, 3, 4/. All of them show common attributes, which are a part of the presented system, too. However, in contrary to these systems, which are general purpose architecture development tools, the present one is dedicated to systolic architectures. That results in a great simplicity in the development of the architecture model and sugnifficant reduction in the corresponding design time. Primary feature of the tool is its capability of symbolic functional analysis, which is not possible for general purpose architecturees of the same complexity. The system is provided also with an architecture simulation capability.

In order to analyse or simulate a certain architecture, one must first create a computer model. The model development accepted here is structured so that it can be done in a more or less mechanical way. In accordance with the two main capabilities of the system - functional analysis and functional simulation - two models of the considered architecture are created. They differ in the data structures used as well as in the data processing. However, these differences are not seen by the user and are inner for the system. The architecture description language is common for the analysis and simulation models. The only statement to be done is an explicite statement

MODE: ANALYSIS

for architecture analysis, or

MODE: SIMULATION

for architecture simulation. With that statement all variables an arrays are set to be of STRING type in case of analysis and of type REAL in the case of architecture simulation. The length of the strings or the width of the reals can be set explicitly in braces after the mode statement.

From this point on, all declarations necessary for the development of the architecture model are one and the same for analysis and simulation. Model creation begins with setting a name for the processing element to be used. After that all inputs and outputs of the processing element (the cell) are listed , followed by the inner registers. An explicit declaration of the type of the variables is necessary, because different data structures in the model correspond to inputs and outputs on the one hand and inner registers on the other hand. That is why, all input names start with X, all outputs - with Y, and all inner registers - with Z.

The description of the function of the cell is carried out in terms of high level programing language used to implement the architecture description language (in this case BASIC). However, all arithmetic and logic expressions are presented as concatenations of the corresponding variables and operators. For example the operation

```
ZC = ZC + ZA.ZB
```

is presented as

```
ZC = ZC + "+" + ZA + "." + ZB.
```

That approach is used to support the primary capability of the program system - symbolic functional analysis. However, the same expression is used in the creation of the model for system simulation. Anyway, in the process of compilation of the model all quotation marks are disregarded, so that the cell function is emulated. The primitive cell of the previously considered example should be described by the following module:

```
CELL: STEP PROCESSOR
      XA, XB, XC
      YA, YB, YC
      ZA, ZB, ZC
FUNCTION:
      ZA = XA
      ZB = XB
      ZC = XC
      ZC = ZC + "+" + ZA + "." + ZB
      YA = ZA
      YB = ZB
      YC = ZC
ENDCELL
```

The cell description is followed by the description of the interconnection pattern. The regularity of the interconnection pattern of the systolic architectures results in a great reduction in the model development effort. Unlike most architecture description languages, in which common interconnection points (nodes) are listed with the ports attached to them, in the present architecture development language the interconnections are described by means of recurencies. In the interconnection description module the dimension of the processor array is also specified:

```
NETS:
        DIM  I = 4, K = 4
        XA(I+1,K) := YA(I,K)
        XB(I+1,K) := YB(I,K)
        XC(I+1,K) := YC(I.K)
ENDNETS
```

In that step data arrays are created and properly dimensioned, so that no dimension violations appear as a result of the interconnection definitions. The operator ":=" does not make the value of the variable at its left equal to the value of the variable at its right, but rather sets them equal to one and the same valme. That is achieved by the special data structure used to present the interconnection pattern: the names XA(I,K), YA(I,K) etc. are identificators of pointers to the address of the value and the operator ":=" sets these pointers equal to one another. That approach saves system resorses and runtime.

If irregularities have to be introduced in the interconnection pattern, that is done in a special NETSI module. The statements in that module have priority with respect to the statements of the module NETS.
After the cell and the interconnection pattern are described, the model is compiled and all free (unsaturated - analogously to valency bonds in chemistry) inputs and outputs are determined and printed out. Normally these inputs and outputs are also input/output ports of the array.

For every clock period the input values are entered at the corresponding input ports of the processor array in the order listed by the system itself. In case of functional analysis these input values are character strings, e. g. a11, a12,..., b11, b12 etc. These character strings are concatenated with oneanother and with the corresponding operators, so that at the output ports concatenations of the input character strings appear, e. g. a11.b11 + a12.b21 etc. The dependance of the output values from the input ones may be made explicit by entering the names of the input ports and the clock period at which the entry is made. In case of simulation numbers are entered instead of character strings and numbers appear at the outputs just as in the real system.

The analysis and the simulation are carried out in an user-friendly environment by providing the system with interactive capabilities. All input values are entered in the order listed by the system and after the entry of the last one for a given clock period, analysis (respectivly simulation) is carried out immediately and the output values are printed out. The state of a sertain inner register or an interconnection path can be monitored by adding it to the output ports by the command PRINT followed by the name of the register encountered.

RESULTS

The presented system is being developed to reduce the efforts in the design and verification of systolic algorithms and architectures. A number of models of systolic architectures were created in a general purpose high level programming language (BASIC, PASCAL) and common features and problems in the process of modelling have emerged. As in any case, in which extensive programming has to be done, it turned out reasonable to develop a dedicated high level language for the description of systolic architectures.

The first version of the presented system worked in a totally interactive environment, and model creation, analysis and simulation were carried out by answering prompts of the system. That version was written in BASIC and run on an APPLE II personal computer. Because of some disadvantages in the editting capabilities of that first version, the presented architecture description language was developed. At present it is being implemented in BASIC and a PASCAL implementation is planned.

Several systolic algorithms and architectures have been tested with different interconnection patterns, cell functions, and data input patterns. Already suggested systolic algorithms have been identified and alternative architectures were found for some of them. New systolic algorithms were created, but they need yet a clear interpretation and a practical application is to be specified. These results will be reported elsewhere.

CONCLUSIONS

It has been shown that traditional problems in lower level design stages as electrical, logical, and physical design, are not so severe in the design of IC implementing systolic algorithms. That makes systolic algorithms very suitable for IC implementation and considerable effort should be given to the design of systolic algorithms and specification of the problems, which can be solved with their help. Adequate design automation tools have to be constructed for that purpose.

The system presented above does not give an algorithm for synthesis of systolic architectures and algorithms. However, it supports the designer in the analysis of a given architecture and varifies its proper function. By keeping an account of what happens in the array in an explicit form and how changes in the interconnection pattern, processor function, or input data flow affect the result, it may give important insights in the problem how new systolic architectures would work. Therefore we consider the presented system for functional analysis and simulation as a helpful tool in the investigation of new systolic algorithms and architectures.

REFERENCES

/1/ H. T. Kung and Ch. E. Leiserson, ch. 8.3 "Algorithms for VLSI Processor Arrays", in "Introduction to VLSI Systems" by C. A. Mead and L. A. Conway, Addison-Wesley, Reading, Mass., 1980.

/2/ M. R. Barbacci and D. Siewiorek, "An Architectural Research Facility - ISP Description, Simulation, Data Collection", AFIPS Conf. Proceedings, Vol. 46, 1977.

/3/ F. I. Parke, "An Introduction to the N.mPc Design Environment", Proc. 16. Design Automation Conf., 1979.

/4/ D. Hill, "SABLE: A Tool for Generating Structured Multi-Level Simulations", Proc. 16. Design Automation Conf., 1979.

OPTIMAL VLSI ALGORITHMS FOR SELECTION THE MAXIMUM ELEMENT OF A SET

Imrich Vrto [1)]

Abstract

We present optimal VLSI algorithms for selection the maximum element of a set. We analyze the behaviour of the algorithms in light of three measures of VLSI complexity: the chip area A, the time of computation T and composite measure AT^2. The analysis is based on computational model of VLSI circuit with major assumptions: semelectivity, the linear propagation of information and convexity of a chip. Is shown that for the problem of selection of maximum number of n integers each of them being represented on k bits holds: $A = \Theta(\min(n,k))$, $T = \Theta((nk)^{1/2})$ and for $k \leq n$ $AT^2 = \Theta((nk)^{3/2})$

1. Introduction

Selection of the maximum element of a set is very frequent computational problem. VLSI technology allows to implement algorithms for this problem on a single chip. In this paper we present optimal VLSI algorithms for selection the maximum element of a set. We analyse the behaviour of the algorithms in light of three measures of VLSI complexity: the chip area A, the time of computation T and composite measure AT^2. The analyses of algorithms is based on realistic computational model of VLSI circuit proposed by Chazelle and Monier [1]. The major assumptions of that model are semelectivity, the linear propagation of information and convexity of a chip. For the above measures of complexity lower bounds of the problem are proved and for each measure an optimal algoritm is designed. It is shown that for the problem of selection of maximum element of n integers $a_1, a_2, \dots, a_n$ each of them being represented on k bits, $a_i = a_{ik}a_{ik-1} \dots a_{i1}$, $i = 1,2,\dots,n$, holds: $A = \Theta(\min(n,k))$, $T = \Theta((nk)^{1/2})$ and for $k \leq n$, $AT^2 = \Theta((nk)^{3/2})$.

This paper is organized in the following fashion. In the next section we define a model of VLSI computation. Section 3 contain our lower bounds for the problem. Our upper bounds appear in section 4. Finaly we close the paper with a summary of our result and a short discussion.

2. Model

Our lower and upper bounds are based on the model of VLSI computation introduced in [1]. We can think of a VLSI circuit as a computatio-

---

1) Institute of Technical Cybernetics of Slovak Academy of Sciences, Dúbravská cesta 9, Bratislava 842 37

nal device which computes a Boolean function $(y_1,y_2,\dots,y_r) = F(x_1,x_2,x_3,\dots,x_s)$. Information is treated digitally and is communicated through I/O ports. Internally, a computational device is a circuit connected nodes and wires into some graph and is defined by its geometrical layout. Actual computation of a problem can be viewed as the propagation and logical treatment of information bits across nodes and wires. Each node represents a finite state machine with constant numer of inputs, outputs and state varibles. All nodes are on the same plane and wires are allowed to lie on a constant number of parallel layers.

A width of wires must exeed a minimum value λ (unit of length). All nodes occupy a fixed area. We distinguish I/O ports where input and output values are available from the logical nodes (gates) which compute Boolean functions. The circuit is laid out within a convex region with all I/O ports lying on its boundary.

A node can produce a result only a delay τ (unit of time) after its inputs have been available. The time of propagation of information across a wire is proportional to the length of the wire. Each input bit is read only once (semelectivity).

Three parameters characterize the computational complexity of a circuit: its area A, time of computation T and composite measure AT^2. The area is the area (in λ^2) of a smallest rectangle covering the geometrical layout of a circuit. T measures the maximal time for all possible inputs, between apparences of the first input bit and the last output bit. The composite measure AT^2 expresses a tradeoff between area and time.

3. Lower bounds

In this section we concentrate upon derivating lower bounds on area, time and area.time^2 product.

3.1 Lower bound on area

Our deriving a lower bound on area is based on the method described by Savage [2]. Let $Y = f(x_1,x_2,\dots,x_s)$ is a Boolean function and p,q are integers. Let J be a subset of $\{1,2,\dots,s\}$. Then the notation $f\Big|_{j\ J}^{x_j=c_j}$ describes the subfunction obtained by substituting the values $x_j=c_j \in \{0,1\}$ for each integer j J. The class $P_{p,q}$ of Boolean functions is defined as $P_{p,q} = \{ f:\{0,1\}^s \to \{0,1\}$ for all $J \in \{1,2,\dots,s\}$, $|J| = p$, there exist constants $c_j \in \{0,1\}$, $j \in J$, such that there are at least q distinct subfunction $f\Big|_{j\ J}^{x_j=j} \}$.

Lemma 1 [2]: Let $f \in P_{p,q}$ be a function computed by semelective VLSI circuit. Then $A = \Omega(\log q)$.

Befor introducing the result on area we will prove the following

Lemma 2: Let us consider a square table of type mxm which contains $m^2/2$ zeros and $m^2/2$ ones. Then at least one of the cases has to happen:

I. There exist two columns U,V with the properties: U contains in some m/6 rows zeros, V contains in the same m/6 rows ones and one of these m/6 rows is the first.

II. There exist two rows U,V with the properties: U is the first row and contains in some m/6 columns the same number x, V contains in the same m/6 columns number $\bar{x}$, where x denote 0 or 1.

Proof: Let us consider the first row of the table. Wlog we can assume that it contains exactly z zeros, $z \geq m/2$ and these zeros lie in the first z columns. If there exists a row containing at least m/6 ones in the first z columns then lemma 2 holds.

Let each row contains less then m/6 ones in the first z columns. Cleary $z \leq 2m/3$, because in the opposite case there would be more than $2m^2/3 - m^2/6 = m^2/2$ zeros in the subtable generated by the first z columns. Further, it is evident that among the first z columns there exists a column U which contain at least $(zm - m^2/6)/z$ zeros.

We can make similar consideration for the last m-z columns with conclusion that there exists column V among the last m-z columns which contain at least $((m-z)m - m^2/6)/(m-z)$ ones. Then column U contains in $(zm - m^2/6)/z + ((m - z)m - m^2/6/(m - z) - m \geq m/6$ rows zeros and V contains in the same rows ones.

Now we are ready to prove the main result.

Theorem 1: Any semelective VLSI circuit selecting the maximum element from n k-bit numbers requires area $A = \Omega(\min(n,k))$.

Proof: It is evident that the problem of selecting the maximum number from n k-bit numbers does not require less area than the problem of computing the least significant bit of the maximum number from m m-bit numbers, where $m = \min(n,k)$. Let the Boolean function $y = f(a_{1m}, a_{1m-1}, \ldots, a_{11}, a_{2m}, \ldots, a_{m1})$ describes this simplyfied problem. We shall show that $f \in P_{p,q}$, where $p = m^2/2$, $q = 2^{m/6} - 1$. According to lemma 1 the assertion of theorem follows immediately.

Let us assume that the variables a_{ij} are placed in mxm table so that ij-th element is equal to a_{ji}, i.e. the i-th column represent the number a_i and j-th row represents the j-th bits of numbers $a_1, a_2, \ldots, a_m$.

Let J is any subset of the set of indices $\{ij \mid 1 \leq i,j \leq m\}$ with property $|J| = m^2/2$. According to lemma 2 we distinguish two cases.

I. There exist two columns U,V in the table of variables with properties: U contains in some m/6 rows variables with indices belonging to J and V contains in the same m/6 rows variables with indices not belonging to J. Wlog we can assume that these m/6 rows are the first.

Now we define $2^{m/6}$ subfunction of f. Let us set the first m/6 variables of the column U to all possible values. All other variables with indices belonging to J let us set to zero. We assert that all subfunction are distinct. Let $f_1(f_2)$ is subfunction of f generated by setting the first m/6 variables of the column U to some m/6-bit number $c_1(c_2)$. Cleary $c_1 \neq c_2$. Let $c_1 < c_2$ and let b is the least significant bit of c_2. Let us set the first m/6 variables of the column V to values so that these values

generate some m/6-bit number d with properties $c_1 \leq d < c_2$ and LSB(d)= $\bar{b}$
All other variables let us set to zero. Evidently the output of f_1 is $\bar{b}$ and the output of f_2 is b.

II.a There exist two rows U,V with properties: U is the first row and contains in some m/6 columns variables with indices not belonging to J. V contains in the same m/6 columns variables with indices belonging to J. Wlog we can assume that these m/6 columns are the first and V is the second row.

Let us set the first m/6 variables of V to all possible values except m/6 zeros. All other variables with indices belonging to J let us set to zero. We obtain $2^{m/6}-1$ distinct subfunction of f. Actually, let f_1, f_2 are two subfunction differing in assignement of values to a variable a_{12}, where $1 \leq 1 \leq m/6$. Let the value of a_{12} of subfunction $f_1(f_2)$ is zero (one). Let us set the variables of f_1, f_2 to the values in the following way: $a_{11} = 1$ andall other variables are equal to zero. The output of f_1 is 0, the output of f_2 is 1.

II.b There exist two rows U,V with properties: U is the first row and contains in some m/6 columns variables with indices belonging to J. V contains in the same columns variables with indices not belonging to J. The proof of this case is similar to the proof of the case II.a.

3.2 Lower bound on time

The following lemma describes the method which we use for deriwing a lower bound on time.

Lemma 3[1]: Let a Boolean function $y = f(x_1, x_2, \ldots, x_s)$ satisfies the condition C: There exist constants $c_1, c_2, \ldots, c_s$ (so called hard input) such that for any i, i = 1,2,...,s:
$f(c_1, c_2, \ldots, c_{i-1}, 0, c_{i+1}, \ldots, c_s) \neq f(c_1, c_2, \ldots, c_{i-1}, 1, c_{i+1}, \ldots, c_s)$. Then any VLSI circuit computing f requires time $T = \Omega(s^{1/2})$.

Now we are ready to prove lower bound on time.

Theorem 2: Any VLSI circuit computing the maximum number from n k-bit requires time $T = \Omega((nk)^{1/2})$.

Proof: It is evident that VLSI circuit computing $\max(a_1, a_2, \ldots, a_n)$ has at least such time complexity as a circuit computing LSB of $\max(a_1, a_2, \ldots, a_n)$ where $a_1 = 00\ldots01$, $a_i = a_{ik}a_{ik-1}\ldots a_{i2}0$, for i = 2,3,...,n. Let us denote this simplified problem by a Boolean function $y = f(a_{2k}, a_{2k-1}, \ldots, a_{22}, a_{3k}, \ldots, a_{n2})$. Cleary, f is a function of (n-1)(k-1) variables and satisfies the condition C of lemma 3 because (0,0,...,0) is its hard input: $f(0,0,\ldots,0,\ldots 0) = 1$, $f(0,0,\ldots,1,\ldots,0) = 0$.

3.3 Lower bound on AT^2.

Theorem 3: Any VLSI circuit selecting maximum number from n k-bit numbers requires $AT^2 = \Omega((nk)^{3/2})$.

Proof: It holds AT = (nk) [3]. Further, we have $T = \Omega((nk)^{1/2})$ from the theorem 2. Combining these results we obtain the assertion.

4. Upper bounds

In this section we propose optimal algorithms which achieve our lower bounds from section 3.

4.1 Upper bound on area

a) If $k \leq n$, then consider the linear array of k comparison-exchange modules, as depicted in Figure 1. for $k = 4$.

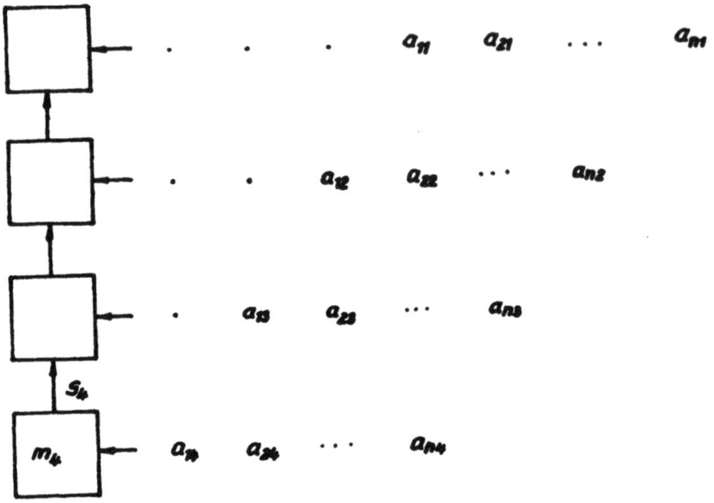

Fig. 1.

The array compare in bit serial fashion the temporal maximum of $(a_1, a_2, \dots, a_{i-1})$ with a_i for $i = 2,3,\dots,n$. The timing of inputs is evident from Fig.1. The i-th modul (from the top to the bottom) has a state variable m_i, which savesthe i-th bit of the temporal maximum. At each step i-th modul compares the value of m_i with some input bit a_{1i} according to information s_{i+1} about comparison of higher bits. The bit of the new temporal maximum is stored in m_i and a new information s_i is generated. Computation begins when a_{1k} enters the k-th modul and is stored in m_k. The comparison-exchange modul can be bilt from O(1) logic gates. Therefore area of array is O(k).

b) Let $k \geq n$. Consider linear array of comparison-exchange modules, as depicted in Fig. 2., for $n = 5$, $k = 6$. The array consists of n-1 modules. Numbers $a_1, a_2, \dots, a_n$ are indicated by vectors and enter the array in bit seriall fashion according to Fig. 2., the most significant bit the first. The first module (from the left to the right) serially compares a_1, a_2 and serially sends $\max(a_1, a_2)$ to the second modul. The second module compare $\max(a_1, a_2)$ with a_3 and so on. Every module has a variable s, which saves an information about comparison of higher bits. The area of the array is O(n) because the area of one module is evidently O(1).

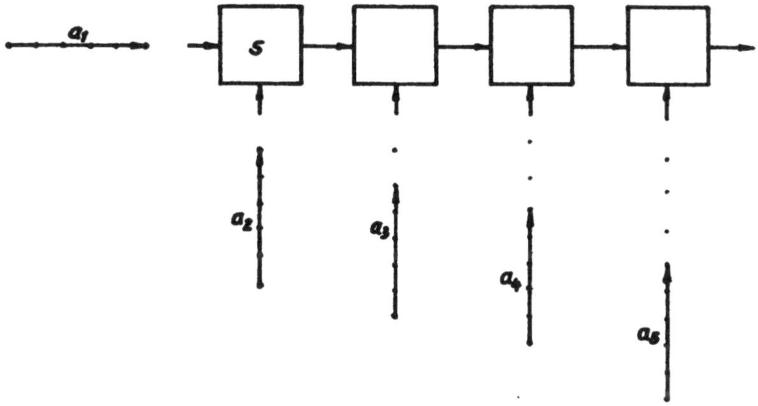

Fig. 2.

4.2 Upper bound on time

First, we describe a basic processor from which we shall construct array for finding the maximum element in $O((nk)^{1/2})$ time. This processor computes the maximum of two k-bit number a_1, a_2 in $O(k^{1/2})$ time and is depicted in Fig. 3., for k = 9. Processor consists of k square and $k^{1/2}$ circular modules. The square modules are arranged into a mesh of type $k^{1/2} \times k^{1/2}$. Computation consists of three stages:

I. Bits of a_1, a_2 are read by processor.

II. In every column of square modules the stored bits are serially compared from more significant bits to less significant bits like in algorithm of 4.1.a. But bits of a_1, a_2 are still saved. The result s_i, i = 1, 2,...,$k^{1/2}$, of comparison in every column of square modules is generated and stored in circullar modules.

III. The result s_1 of comparison of highest $k^{1/2}$ bits is sent to the left. According to value of s_1 the new value of s_2 is computed and this process continues to the left. Finally, the values of s_i contain the right information about max(a_1, a_2). Then the values of s_i are sent back to the square modules and the bits of min(a_1, a_2) are cleaned.

Evidently, every stage requires $O(k^{1/2})$ time. It is easy to generalize the above computation for selection the maximum element from l numbers in time $O(lk^{1/2})$.

Now consider a linear array of $n^{1/2}$ basic processors. The processors simultaneously compute the maximum from $n^{1/2}$ numbers in $O\ (nk)^{1/2}$ time.

Then the processors serially compute the maximum from those $n^{1/2}$ temporal maximum numbers in $O((nk)^{1/2})$ time units.

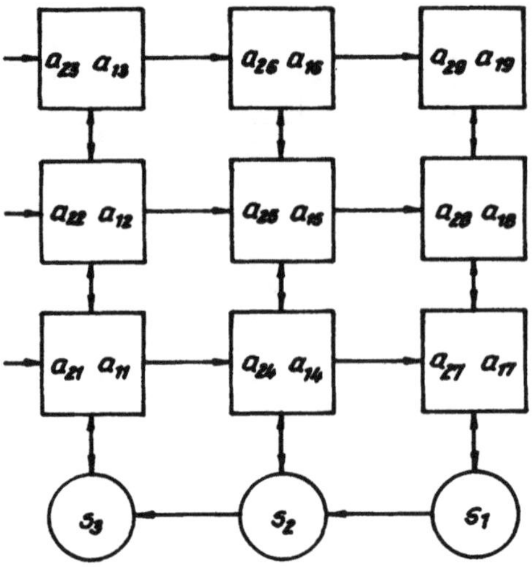

Fig. 3.

4.3 Upper bound on AT^2

In this subsection we propose an optimal algorithm on AT^2 measure for $k \leq n$. Our algorithm is a generalization of the algorithm of 4.1.b). Let $r = (nk)^{1/2}$. Consider a linear array, as depicted in Fig.4. The array consists of r processors of the same type as in the array 4.1.b).

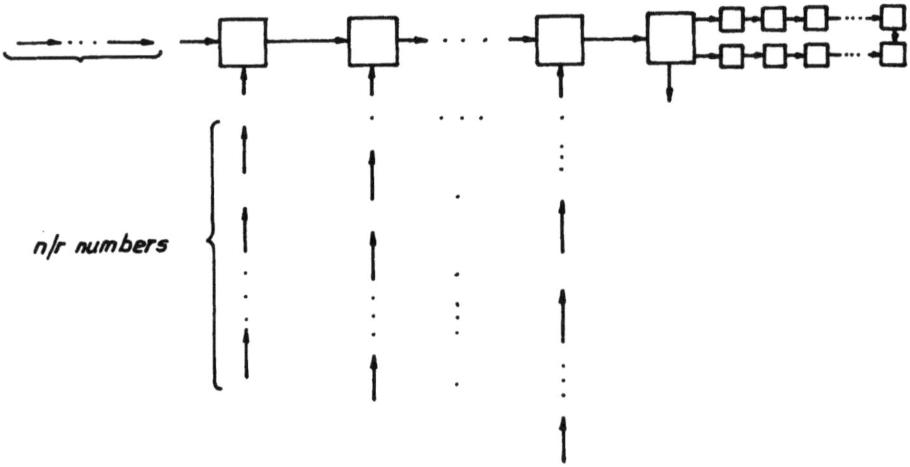

Fig. 4.

The input and timing of numbers (vectors) is evident from the Fig.4. The first r-1 processors (from the left to the right) compute the maximum number of the first r numbers. This maximum is stored in k-bit shift register attached to the r-th processor. Then the maximum of the second r numbers is computed and compared with the maximum from shift register on the r-th processor. The new temporal maximum element is evaluated and stored again in the shift register and s.o.

The computation is fully pipelined therefore this array works in $T = O(nk/r + r) = O((nk)^{1/2})$ time units. The area of the array is $A=O(r)$. Hence $AT^2 = O((nk)^{3/2})$.

5. Conclusions

We have determined the area, time and AT^2 complexity of VLSI circuits for selection the maximum element from n k-bit numbers. Our lower bounds are based on the method of [1], [2]. These methods require semelectivity, linear propagation of information across wires and convexity of a chip with I/O ports lying on its boundary. Note, that they do not need when and where oblvious input and output.

One interesting open question remains: What is the area.time^2 complexity for $k > n$?

References

1 Chazelle, B. and Monier, L.: A Model of Computation for VLSI with Related Complexity Results. In: Proc of 13-th Annual ACM Symposium on Theory of Computing, 1981, pp. 318-325

2 Savage, J. E.: Planar-circuit Complexity and the Performans of VLSI Algorithms. In: VLSI System and Computations. Computer Science Press, 1981, pp. 269-277.

3 Ullman, J.: Computational Aspects of VLSI. Computer Science Press, Rockville Mar., 1984.

Band 1
Inverse and Improperly Posed Problems in Differential Equations
Proceedings of the Conference on Mathematical and Numerical Methods, held in Halle/Saale (GDR) from May 29 to June 2, 1979
Edited by Gottfried Anger
296 Seiten, 1979 (vergriffen)

Band 2
Fundamentals of Computation Theory FCT '79
Proceedings of the Conference on Algebraic, Arithmetic and Categorial Methods in Computation Theory, held in Berlin/Wendisch-Rietz (GDR) September 17—21, 1979
Edited by Lothar Budach
576 Seiten, 1979

Band 3
Decidability and Generalized Quantifiers
By Andreas Baudisch, Detlef Seese, Hans-Peter Tuschik, and Martin Weese
XII, 235 Seiten, 1980

Band 4
Zur Elastizitäts- und Thermoelastizitätstheorie
Potentialtheoretische Methoden in Anwendung auf verschiedene Modelle
Von Lothar Jentsch und Johannes Maul
157 Seiten, 1980

Band 5
Systems Analysis and Simulation 1980
Proceedings of the International Symposium held at Berlin (GDR) September 1—5, 1980
Edited by Achim Sydow
496 Seiten, 1980 (vergriffen)

Band 6
Nichtlineare diskrete Optimierung
Von Eberhard Girlich und Michael M. Kowaljow
218 Seiten, 1981

Band 7
Completeness Problems in the Structural Theory of Automata
By Jürgen Dassow
148 Seiten, 1981

Band 8
Darstellung von Lösungen linearer elliptischer Differentialgleichungen
Von Günther Wildenhain
92 Seiten, 1981

Band 9
Introduction to the Spectral Theory of Operators in Spaces with an Indefinite Metric
By I. S. Iohvidov, M. G. Krein, and H. Langer
120 Seiten, 1982 (vergriffen)

Band 10
Nonlinear Time-discrete Systems
A General Approach by Nonlinear Superposition
By Michael Gössel
115 Seiten, 1982 (vergriffen)

Band 11
Preprimal Algebras
By Klaus Denecke
162 Seiten, 1982

Band 12
Ergodic Theory and Related Topics
Proceedings of the Conference held in Vitte/Hiddensee (GDR)
October 19—23, 1981
Edited by Horst Michel
232 Seiten, 1982

Band 13
Singularly Perturbed Differential Equations
By H. Goering, A. Felgenhauer, G. Lube, H.-G. Roos, and L. Tobiska
176 Seiten, 1983

Band 14
Leben wir in einer Volterra-Welt?
Ein ökologischer Zugang zur angewandten Systemanalyse
Von Manfed Peschel und Werner Mende
233 Seiten, 1983

Band 15
Markov Decision Problems with Countable State Spaces Optimality Criteria — Algorithms — Clustering
By Hans Michael Dietz and Volker Nollau
174 Seiten, 1983

Band 16
Stochastische Verfahren zur Suche nach dem Minimum einer Funktion
Von Ryszard Zieliński und Peter Neumann
133 Seiten, 1983

Band 17
Invertibility and Asymptotics of Toeplitz Matrices
By Albrecht Böttcher and Bernd Silbermann
200 Seiten, 1983

Band 18

Structural Induction on Partial Algebras

Introduction to Theory and Application of Partial Algebras — Part II

By Horst Reichel

205 pages, 1984

Band 19

Algebraic Methods for Toeplitz-like Matrices and Operators

By Georg Heinig and Karla Rost

212 pages, 1984

Band 20

Frege Conference 1984

Proceedings of the International Conference held at Schwerin (GDR), September 10—14, 1984

Edited by Gerd Wechsung

408 pages, 1984

Band 21

Numerical Analysis of Selected Semilinear Differential Equations

Edited by Christian Großmann, Thomas Riedrich, and Hartmut Schönheinz

203 pages, 1984

Band 22

Singular Ordinary Differential Operators and Pseudodifferential Equations

By Johannes Elschner

200 pages, 1985

Band 23

Lotka-Volterra-Approach to Cooperation and Competition in Dynamic Systems

Proceedings of the 5th Meeting of UNESCO's Working Group on Systems Theory held on the Wartburg, Eisenach (GDR) at March 5—9, 1984

Edited by Werner Ebeling and Manfred Peschel

271 pages, 1985

Band 24

Convergence Structures 1984

Proceedings of the International Conference on Convergence Structures and Applications held in Bechyné (Czechoslovakia), September 17—21, 1984

Edited by Josef Novák, Werner Gähler, Horst Herrlich, and Jan Mikusiński

250 pages, 1985

Band 25

Parcella '84

Proceedings of the International Workshop on Parallel Processing by Cellular Automata and Arrays held in Berlin (GDR), September 25—27, 1984

Edited by Wolfgang Händler, Tamás Legendi, and Gottfried Wolf

239 pages, 1985

Band 26

Multiobjective and Stochastic Optimization Based on Parametric Optimization

By Jürgen Guddat, Francisco Guerra Vasquez, Klaus Tammer, and Klaus Wendler

171 pages, 1985

Band 27

Systems Analysis and Simulation 1985 I: Theory and Foundations

Proceedings of the 2nd International Symposium on Systems Analysis and Simulation held in Berlin (GDR), August 26—31, 1985

Edited by Achim Sydow, Manfred Thoma, and Robert Vichnevetsky

496 pages, 1985

Band 28

Systems Analysis and Simulation 1985 II: Applications

Proceedings of the 2nd International Symposium on Systems Analysis and Simulation held in Berlin (GDR), August 26—31, 1985

Edited by Achim Sydow, Manfred Thoma, and Robert Vichnevetsky

496 pages, 1985

Band 29

Pseudo-differential Operators and Mellin Operators in Spaces of Functions with Conormal Singularity

By Stephan Rempel and Bert-Wolfgang Schulze

about 480 pages, 1986

This volume contains the papers of PARCELLA '84 — the II. International Workshop "Parallel Processing by Cellular Automata and Arrays" held in Berlin (GDR) from September 25 to 27, 1984.
This workshop united those scientists who were interested in discussing the new results and ideas for parallel and distributed processing on homogeneous computer structures. Cellular systems and array systems are main representatives of this structure type.
The existence side-by-side and the competition between these two closely related concepts appears to be fruitful for their own development as well as for the overall development of further computer architecture in general.

Dieser Band enthält die Beiträge von PARCELLA '84 — dem II. Internationalen Workshop „Parallel Processing by Cellular Automata and Arrays", das vom 25. bis 27. September 1984 in Berlin (DDR) stattfand.
Dieses Workshop hat Wissenschaftler zusammengeführt, die an der Diskussion neuer Ergebnisse und Ideen auf dem Gebiet der parallelen und der verteilten Verarbeitung auf homogenen Rechnerstrukturen interessiert sind.
Zellulare Systeme und Arraysysteme sind die Hauptvertreter dieses Strukturtyps.
Aus dem Nebeneinander und dem Wettstreit dieser zwei eng miteinander verwandten Konzepte werden Impulse sowohl für die Entwicklung dieser Konzepte als auch über diesen Bereich hinaus für die Entwicklung zukünftiger Rechenarchitekturen insgesamt erwartet.

ISSN 0138-3019

Bei Fragen zur Produktsicherheit wenden Sie sich bitte an:
If you have any questions regarding product safety,
please contact:

Walter de Gruyter GmbH
Genthiner Straße 13
10785 Berlin
productsafety@degruyterbrill.com